Baldwin Spencer, William Austin Horn

Report on the Work of the Horn Scientific Expedition to Central Australia

Vol. 1

Baldwin Spencer, William Austin Horn

Report on the Work of the Horn Scientific Expedition to Central Australia
Vol. 1

ISBN/EAN: 9783337329983

Printed in Europe, USA, Canada, Australia, Japan

Cover: Foto ©Andreas Hilbeck / pixelio.de

More available books at **www.hansebooks.com**

REPORT ON THE WORK

OF THE

HORN SCIENTIFIC EXPEDITION

TO

CENTRAL AUSTRALIA.

PART I.—INTRODUCTION, NARRATIVE, SUMMARY OF RESULTS, SUPPLEMENT TO ZOOLOGICAL REPORT, MAP.

EDITED BY

BALDWIN SPENCER, M.A.,
PROFESSOR OF BIOLOGY IN THE UNIVERSITY OF MELBOURNE.

London
DULAU AND CO., 37 SOHO SQUARE.

Melbourne
MELVILLE, MULLEN AND SLADE.

SEPTEMBER, 1896.

PART I. NARRATIVE AND SUMMARY OF RESULTS.

TABLE OF CONTENTS.

	PAGE
INTRODUCTION, by W. A. HORN	v.
EDITORIAL NOTE	xi.
THROUGH LARAPINTA LAND; A NARRATIVE of the Expedition, by BALDWIN SPENCER, M.A., C.M.Z.S.	1
SUMMARY of the Zoological, Botanical, and Geological Results of the Expedition, by BALDWIN SPENCER, M.A., C.M.Z.S.	139
ZOOLOGY	139
BOTANY	159
GEOLOGY AND PALÆONTOLOGY	162
GENERAL CONCLUSIONS	171
SUPPLEMENT to the Zoological Report	201
HYMENOPTERA, by W. F. KIRBY, F.L.S., F.E.S.	203
ADDITIONS TO THE FAUNA	210
INDEX	212
MAP.	

INTRODUCTION.

THE scientific exploration of Central Australia, more particularly that portion known as the McDonnell Ranges, had for many years been desired by the leading scientific men in Australia, some of whom hold the opinion that when the rest of the continent was submerged the elevated portions of the McDonnell Range existed as an island, and that consequently older forms of life might be found in the more inaccessible parts. Travellers' tales also of the manners and customs of the natives, and the varieties of plants and animal life in these remote regions, had aroused a widespread interest, and at the solicitation of a few scientific friends I resolved to organise and equip a party, composed of scientific men, to thoroughly explore this belt of country. The proposition was received with great favour in Australia, and numerous applications were made, and even premiums offered, by gentlemen anxious to join the Expedition. The failure, however, of previous Expeditions made it necessary to exercise great care in the selection of the various members, so as to avoid the disasters, in the shape of internal dissensions, which had wrecked the others. In order to secure the services of the best men in Australia I decided to make it a semi-national undertaking, and to this end invitations were extended to the Premiers of the principal Colonies, asking them to nominate scientific representatives.

The Premiers of the Colonies of Victoria, New South Wales, and South Australia responded most cordially. Victoria, partly through the generous influence of Lord Hopetoun, nominated Professor Baldwin Spencer, of the Melbourne University. New South Wales nominated Mr. J. Alexander Watt, of the Sydney University, and South Australia nominated Professor Ralph Tate, F.L.S., and Dr. Edward Stirling, F.R.S., both of the Adelaide University. Mr. C. A. Winnecke, F.R.G.S., was chosen as the surveyor and meteorologist, and the fact that, in addition to piloting the party to such points as they wished to visit, this gentleman traversed and plotted about 27,000 square miles of country, and also made a series of valuable meteorological observations, speaks for itself.

The general public were for some time under the impression that the Expedition was going out in search of gold. They could not understand a body of scientific gentlemen going into a desert country, giving up their time and services, and submitting to all the dangers, discomforts and hardships attendant upon the life for any other reason. There is no doubt that had one of the collectors in pursuit of a butterfly barked his shins against a nugget of gold, he would have recognised, and having recognised, would have "collected" it, although his claim would probably have been disputed by the geological section of the Expedition. But the real objects of the Expedition were as set out in the articles under which they started, viz., the scientific examination of the country from Oodnadatta to the McDonnell Range; the collection of specimens illustrative of the fauna, flora, and geological structure and mineralogical resources of that region, and the illustration by photography of any remarkable natural features of the country traversed; the securing of photographs of the aborigines in their primitive state, the collection of information as to their manners, customs, and language, and the reproduction of their mural paintings.

We made our final start from Oodnadatta, which is the northern terminal point of the railway from Adelaide, on 6th May, 1894. Our party consisted of, in addition to the scientific gentlemen already named, two Afghan and two European camel-drivers, two collectors, two prospectors, one aboriginal black tracker, and one cook, making sixteen in all, with twenty-six camels and two horses. Without pretending to any great amount of scientific knowledge myself, I have had considerable experience in bush life, extending over many years, and had done a good deal of exploring work in the Eremian region; and, at the solicitation of several members of the party, I accompanied them to a point 1000 miles north of Adelaide, and, finding that they were all working together with the utmost harmony and enthusiasm, I started on my lonely return journey. When leaving I tried the new experiment of having no autocratic leader, but gave each scientific member of the party one vote, so that all questions as to the route to be taken, the length of time to be spent at one spot, or any kindred questions, were decided by the majority. The safe-conduct of the party to such points as they wished to visit was entrusted to Mr. Winnecke.

The continent of Australia extends from the 38th to the 12th parallel of S. Lat., and from the 113th to 153rd degree of Longitude. Now, if we take Ayers Rock as the centre of an ellipse which has a length of 1,600 miles by a width of 800 miles, we have an area which comprises practically the whole of this Eremian region, which has an average rainfall of from five to twelve inches; but this rain fall is very irregular, as long periods of drought, sometimes of two years' duration, frequently intervene, and much of the country is reduced to the condition of an almost impassable desert, thus rendering the close examination of the central portion a task of no small difficulty and occasional danger, firstly from the scarcity of permanent water, and secondly from the presence of occasionally hostile natives.

INTRODUCTION.

I was particularly anxious to obtain, if possible, photographs of Mount Olga and Ayers Rock, which have been regarded as two of the most striking natural features in the central region. As these lie far out in the desert country to the south of Lake Amadeus, it was impossible for the whole party to visit them; but under the leadership of Mounted Trooper E. C. Cowle, to whom my thanks are due for the assistance which he rendered to us, a small party was enabled to pay a flying visit to them, and reproductions of photographs of these striking features, taken by Professor Spencer, appear in the Narrative.

In the very centre of the continent, and within the limits of the Eremian region, there exists an elevated tract of country, known as the McDonnell Ranges. These mountains, barren and rugged in the extreme, rise to an altitude of nearly 5,000 feet above sea-level, while the country surrounding them has an elevation of about 2,000 feet above sea-level; it slopes away on every side towards the coast, distant 1,000 miles. The mountains are at the head of the River Finke, and for this region, including the valley of the Finke, we have adopted the name of Larapintine, from the native name of the Finke, "Larapinta," and it was over this area that most of our explorations were conducted. The existence of these mountains has to a great extent redeemed this portion of the continent from becoming an absolute desert, as the mountains attract the tropical clouds, and during the occasional heavy downpours of rain a vast amount of storm water rushes down their barren rocky sides into the channel of the Finke River and its tributaries, and overflowing the banks inundates a great deal of the surrounding country, particularly in the south. The consequence of such inundation is that over the flooded portion of the country, and also other lowlands on which the rain has fallen, there is a rapid and luxuriant growth of vegetation. The ground being warm the rapidity of the vegetable growth is almost marvellous. I have seen portions of this Eremian region which have been reduced by drought to the condition of a moving mass of sand, and yet within a month of a heavy fall of rain, the country was covered with a most luxuriant vegetation and capable of carrying an enormous amount of stock. These rapid changes have, however, led to ruinous losses among the pastoralists, as people with a meagre knowledge of the climate, and who have seen this country for the first time after one of those tropical downpours, imagine it to be its normal condition, and are induced to send out large numbers of stock to graze; and when the inevitable drought occurs and the country is again reduced to the desert condition, they find their stock dying by hundreds of thousands for want of water.

The climate of the McDonnells in winter is simply perfect, with warm clear days and bright cold nights. Day succeeds day without a cloud. In the afternoon there is generally a light breeze from the S or S.E. The result of observations taken on eighty-four days shows that on twenty-six days a dead calm prevailed; on thirty two days a gentle S.E. wind; on fifteen days a S. or E. wind; on eleven days wind N.W. or S.W.

In South Australia the hot winds are invariably from the north, and this gave rise to the theory that the winds became heated from passing over the dry hot centre of the continent; but hot winds in the centre are much rarer than in the south. During nearly four months there was not enough rain to wet a pocket handkerchief, and it was never necessary to erect the tents. We always slept in the open air.

Climatic conditions have a marked influence on the animal life indigenous to these regions, and have led to the occurrence of some strange phenomena, which are dealt with in the Zoological Report.

From the number of fossil diprotodonts of gigantic size and struthious birds rivalling in stature the New Zealand moa, which have been found within the limits of the Eremian region, it is evident that it had at one time a far heavier and more constant rainfall and a more luxuriant vegetation, capable of sustaining larger and slower-moving forms of animal life than at present. At Lake Callabonna, in the great salt Lake Eyre basin, there are hundreds of fossil skeletons of these animals, some of which have been successfully removed to the Adelaide Museum. In that locality they are found most frequently on the surface of the dry salt lake, and have been preserved by a natural coating of carbonate of lime; but I have found their bones at a depth of twelve feet from the surface, at a place 600 miles S.E. of the McDonnell Range.

I have always felt that it was the duty of some one to obtain accurate information as to the manners, customs, superstitions, etc., of the primitive races which inhabited the continent of Australia before the advent of Europeans, and also to obtain by photography some faithful reproductions of their ceremonial dresses and general appearance before they had come under the debasing influences of the white man. And in this matter we were most ably and generously assisted by Mr F. J. Gillen, who has had a long experience among them and is himself an expert photographer. The race is fast dying out, and there are very few tribes left in their primitive condition who have not been in contact with Europeans; there are all confined to the Eremian region. In this matter, thanks to the assistance of Mr. Gillen, we have been signally successful, and have obtained a very large number of valuable photographs, some of them being of ceremonies and rites which are very rarely witnessed by white men, and have also obtained a mass of reliable information as to their superstitions and general customs, copies of a number of their mural paintings, and a very large collection of their weapons and instruments.

The Central Australian aborigine is the living representative of a stone age, who still fashions his spear-heads and knives from flint or sandstone and performs the most daring surgical operations with them. His origin and history are lost in the gloomy mists of the past. He has no written records and few oral traditions. In appearance he is a naked, hirsute savage, with a type of features occasionally pronouncedly Jewish. He is by nature light-hearted, merry and prone to laughter, a splendid mimic, supple-jointed, with an unerring hand that works in perfect unison with his eye, which is as keen as that of an eagle. He has never been known to wash. He has no private ownership of land, except as regards that which is not over carefully concealed about his person. He cultivates nothing, but lives entirely on the spoils of the chase, and although the thermometer frequently ranges from 15 degrees to over 90 degrees F. in twenty-four hours, and his country is by no means devoid of furred game, he makes no use of the skins for clothing, but goes about during the day and sleeps in the open at night perfectly nude. He builds no permanent habitation and usually camps where night or fatigue overtakes him.

He can travel from point to point for hundreds of miles through the pathless bush with unerring precision, and can track an animal over rocks and stones, where a European eye would be unable to distinguish a mark. He is a keen observer and knows the habits and changes of form of every variety of animal or vegetable life in his country. Religious belief he has none, but is excessively superstitious, living in constant dread of an Evil Spirit which is supposed to lurk round his camp at night. He has no gratitude except that of the anticipatory order, and is as treacherous as Judas. He has no traditions, and yet continues to practise with scrupulous exactness a number of hideous customs and ceremonies which have been handed from his fathers, and of the origin or reason of which he knows nothing. Oft-times kind and even affectionate to those of his children who have been permitted to live, he yet practises, without any reason except that his father did so before him, the most cruel and revolting mutilations upon the young men and maidens of his tribe.

Yet withal he is a philosopher who accepts feast or famine without a murmur either at the pangs of hunger or the discomforts of repletion. His motto is "*Carpe diem,*" and when fortune sends him a supply of game he consumes it all, regardless of to-morrow. No cold missionary graces his side-board, and should hunger, as a penalty for his improvident gluttony, overtake him, he simply ties a thin hair girdle tightly round his stomach, and almost persuades himself that he is still suffering from repletion. After an experience of many years I say without hesitation that he is absolutely untameable. You may clothe and care for him for years, when suddenly the demon of unrest takes possession; he throws off all his clothing and plunges into the trackless depths of his native bush, at once reverting to his old and hideous customs, and when sated, after months of privation, he will return again to clothing and civilisation, only to repeat the performance later on. Verily his moods are as eccentric as the flight of his own boomerang. Thanks to the untiring efforts of the missionary and the stockman, he is being rapidly "civilised" off the face of the earth, and in another hundred years the sole remaining evidence of his existence will be the fragments of flint which he has fashioned so rudely. It was for this reason that I thought it desirable to get some reliable information, supplemented by photography, of this race while there were any of them remaining in their primitive condition.

In order to bring the scientific results together and to make them available, in what appeared to be the most convenient way, to those interested in the various branches of work, they have been published in book form under the editorship of Professor Baldwin Spencer.

To the South Australian Government my thanks are due for the cordial assistance rendered to the Expedition in various ways, especially in the loan of camels; to the Governments of Victoria and New South Wales for their assistance in the nomination of members of the scientific staff to represent those colonies; and to the Councils of the Universities of Adelaide and Melbourne for readily granting to Professor Tate, Dr. Stirling and Professor Spencer the necessary leave of absence.

W. A. HORN.

London, 1896.

EDITORIAL NOTE.

On the return of the Expedition in August, 1894, some little time elapsed before all the material collected reached Adelaide and Melbourne and could be distributed to specialists. During the two years which have since passed by, the working out of the material, writing of reports and reproduction of illustrations has been proceeded with as rapidly as possible.

The Zoological collection has been largely increased by the cordial co-operation of Mr. P. M. Byrne of Charlotte Waters, Mr. F. J. Gillen of Alice Springs, Mr. E. C. Cowle of Illamurta, and Messrs. P. Squires, J. Field and J. Besley of Alice Springs.

To Mr. P. M. Byrne we are especially indebted for the opportunity of securing and describing a number of interesting forms, amongst which are no fewer than four new species of Marsupials captured in the neighbourhood of Charlotte Waters.

Mr. F. J. Gillen generously placed his valuable anthropological notes at the disposal of Dr. Stirling, and they appear as a special article in the Anthropological Section, which is also largely illustrated by reproductions of photographs taken and lent by Mr. Gillen for the purpose.

The map has been compiled and the reproduction of photographs has been executed in London under the personal supervision of Mr. Horn ; the remainder of the illustrations have been lithographed and all the letterpress printed in Melbourne under my own supervision and I have to express my thanks to Mr. Wendel for the skill which he has displayed and the trouble which he has taken in rendering the lithographs as accurate as possible. The work has been issued in parts both to facilitate publication and to allow of the portions dealing with separate branches being more accessible to workers than they would perhaps have been had the volume been published as a whole.

In the work of editing I have to thank my colleagues on the Expedition, Professor Tate, Dr. Stirling, and Mr. Watt for their cordial co-operation, and I am also much indebted to Mr. Winnecke for valuable information, and to Mr. T. S. Hall, Demonstrator and Lecturer on Biology in the Melbourne University, both for assistance in the preparation of the indexes and for suggestions and help in many ways.

Throughout our Expedition everything was done both by official and private individuals with whom we came in contact to render it as successful as possible.

In conclusion the hope may be expressed that our work may form a contribution of some importance to a better knowledge of the great interior of the continent in regard to the various branches of science with which it deals, and may also be regarded as justifying the public spirited enterprise and efforts of its promoter.

BALDWIN SPENCER.

UNIVERSITY OF MELBOURNE,
September, 1896.

Through Larapinta Land

A NARRATIVE OF THE HORN EXPEDITION

TO

CENTRAL AUSTRALIA

By BALDWIN SPENCER, M.A., C.M.Z.S., Professor of Biology in the University of Melbourne.

CONTENTS.

CHAPTER I.—Introductory Remarks.

Object of the Work—Members of the Expedition—Larapinta Land—Difficulties of Travel and Nature of Camels—Departure from Adelaide and Arrival at Oodnadatta—Departure from Oodnadatta—Loading and Riding Camels—Daily Programme while on the March—The Main Sections of the Journey—The Australian Steppes . . . . . *Page* 1

CHAPTER II.—The Lower Steppes.

From Oodnadatta to Charlotte Waters and the Finke River.

Lake Eyre in the Dry and Wet Seasons—Gibber Plains—Origin of the Gibbers—Loamy Plains—The Valley of the Macumba River—Water Holes—Chestnut-eared Finches—The Prickly Seed Cases of Tribulus and Bassia—Succulent Plants, Claytonia and Portulaca—Remarks on Spinous and Succulent Development of Plants—Both forms of growth are probably adaptations to climatic environment and not in the first instance developed as protection against animals—The most spiny and the most succulent plants are found in the arid regions—The Stevenson River—Contents of a Water-hole during the Dry Season—Tenacity of life of Bithinia australis—Dalhousie Station and Mound Springs—Red Mulga—Gibber Plains at Sunset—Clay Pans; contrast between them in the Dry and Wet Seasons—The Fauna of a Clay Pan—Amphibia, Crustacea, Mollusca—Colour Changes of Frogs—Habits of Apus—Fresh Water Crab—Water Holding and Burrowing Frog—The Adminga Creek—Gidden Scrub—Charlotte Waters Telegraph Station—A Second Visit to Charlotte Waters in Summer Time—Flies and Mosquitoes—Succession of Forms of Life—The Colouration of Lizards—Sexual Differences—Brilliant Colouration, the accompaniment of a general state of activity and only indirectly associated with that of the environment—Susceptibility to Heat of Lizards—Tiliqua occipitalis killed by Heat of Sand—Departure from Charlotte Waters—Change in Nature of the Country—Ant Lions—Mount Daniel—Camp at the Goyder River—Habits of Physignathus longirostris—Messrs. Watt and Winnecke start off to follow up the Goyder and Lilla Creeks—The Main Party goes on to Crown Point—View of the Finke Valley . *Page* 11

CHAPTER III.—The Lower Steppes.

From the Finke River to the James Range.

Discovery and naming of the Finke by McDouall Stuart, in 1860—View of the Finke Valley—Cunningham Gap and Crown Point—Camp of Blacks—Their life in Camp—Corroborees—Two important forms, ordinary and sacred—Churiña, sacred Stones and Sticks—Organisation of the Tribe—The way in which they prepare for an ordinary Corroboree—Usual Ornaments, Weapons, and Implements—Women Mourning—Collecting amongst the Sandhills—Pyrameis kershawi and Danais petilia—Scorpions—Deaf Adder—Occurrence and Habits of Limnodynastes ornatus—Two Types of Burrowing Frogs in Central Australia—Departure from Crown Point—Reach the Lilla Creek—Meet Messrs. Watt and Winnecke at the Horse Shoe Bend on the Finke—The Horn Range—Social Caterpillar Cases on Eucalyptus microtheca and Acacias—Various case Moths—Description of the Scrub—Camp at Idracowra—Determine upon Future Plans—Return of Mr. Horn to Adelaide—Visit to Chambers Pillar—Sandhills—Desert Oaks—Description of the Pillar—Myth of the Blacks to account for the

CONTENTS.

Pillar Natural Formation of Water-holes along the Rivers—Sudden appearance of Floods in parts where no Rain has fallen—Presence of Fish in the Waterhole—No Fish in Central Australia known to have taken on the habit of Protopterus, the Mud Fish—Notoryctes typhlops, the Marsupial Mole—Is Notoryctes a form specially modified since climatic conditions became changed in the Central area, or is it the remnant of a once more widely dispersed form?—Departure from Idracowra—Cross the Palmer River and reach Henbury—Waterpool at Henbury—The Bony Bream, Chatoessus horni—Chandler Range and the Ceremonial Stone, Antiarra—Collecting amongst the Blacks—Camped at Henbury—Leave Henbury—Eucalyptus gamophylla—Large Spider Webs in the Scrub—Running Waters on the Finke—Fresh Water Crayfish—Reach Illamurta in the James Range and pass out of the Desert Sandstone Area. *Page*

CHAPTER IV.—The Higher Steppes

The Southern Part of the James Range and the George Gill and Levi Ranges.

The James Range—The Police Camp at Illamurta—Collecting amongst the Ranges—First appearance of Black Earth—Earthworms—Significance of the presence of Acanthodrilus and Microplana in Central Australia—The Ilpilla Creek—Persistence of Land Mollusca amongst the Ranges—Fish in the Water-pools in the Ilpilla Gorge—Absence in Central Australia of anything like a great Mountain Range with sheltered and fertile Valleys—Necessity of being in the district during the various Seasons—Leave Illamurta and travel on to the Palmer River—Camp near to the Illara Waterhole—Native Tobacco Plant—Absence of Frogs and other animals probably due to low temperature at nights—The Party divides into two sections, one going to Tempe Downs the other to the Petermann Creek—Tempe Downs Station—View from the Station Range—The Walker River and Gorge—The habits of the Porcupine-grass Ant—A Corroboree at Tempe Downs—Musical Instruments amongst the Blacks—The Main Camp at the Petermann Creek—Traverse of the Levi Range by Mr. Watt—From the Camp on the Petermann to Trickett Creek and along the southern face of the George Gill Range to Bagot Creek—Our Camps at Bagot and Reedy Creeks—Description of the Reedy Creek Camp—Gum Creeks—View from the Escarpment of the George Gill Range—Collecting amongst the Sandhills to the south of the Range—Jerboa-Rats, Mice and Antechinomys—Tracking of Emus by the Blacks—Penny Springs—Cycads, Encephalartos Macdonnelli—A Picturesque Gorge—Native Rock Drawings at Reedy Creek—Pigments used by the Natives—Division into Two Parties—The Main Camp travels eastward to Laurie Creek and then to the Mcdonnell Range—A Small Party under the guidance of Mr. E. C. Cowle goes south across Lake Amadeus to visit Ayers Rock and Mount Olga. *Page*

CHAPTER V.—The Desert Country.

From the George Gill Range to Ayers Rock and Mount Olga.

Our Equipment—Photographing in Central Australia—Departure from Reedy Creek—Camp for the Night after travelling sixteen miles—Sandhill Gum Trees—Winnall's Ridge, the most Southern Outcrop seen of Silurian Quartzite—The Pituri Plant—Uses to which it is put by the Blacks—Kamaran's Well—A most unlikely Spot for Water—The Remains of a broken down Mound Spring—Dingoes in the Water—Reach Lake Amadeus at Sunset—Cross the Salt Bed and Camp on the South Side—The Present State of Desiccation of the Lake Amadeus Area—Leave Lake Amadeus—Coulthard's Well—Travel all Day over Porcupine Sandhills and in the Afternoon Reach Ayers Rock—View of the Rock from the Sandhills—Camp by a Small Water-hole in a Chasm in the Rock—No Permanent Water at Ayers Rock—Spend the Day round the Rock—Native Drawings on the Walls of Small Caves—Honey Ants—Tadpoles of Helioporus pictus in the Water-hole—View across the Plains towards Mount Olga at Sunset—A Family of Sandhill Blacks—Ride across the Plain to Mount Olga—Camp at the Entrance to a Deep Ravine—Tietkcn's Marked Trees—No Permanent Water at Mount Olga, only a Small Rock-Pool now remaining—Camp of Wild Blacks—Ride back to Ayers Rock—Cooking of a Kangaroo by the Blacks Return to the George Gill Range—Increase of the Water in Bagot Creek—Crossing the George Gill Range—Petermann Pound—Cross the Station Range and reach Tempe Downs Leave Tempe Downs and follow the Walker back to the Palmer—The Gorges along the Palmer—Low Temperature at Night Time—A Large Tussock of Porcupine Grass—Follow the Palmer up to the Missionary Plain and Camp close to Pine Point—A New Species of Grass Tree—Sporadic Distribution of Certain Species of Plants—The Missionary Plains—Gosse Range—Rock Pigeons—Camp in the Southern McDonnell Hills—In the Morning join the Main Party at the Old Glen Helen Station . . *Page* 81

CHAPTER VI.—The Higher Steppes.

The McDonnell Ranges.

Camp at the Base of Mount Sonder—The Redbank Creek and Gorge—Description of Fish found in the Water-holes—The Horn Valley—Origin of the Gorges—Camp in the Finke Gorge—The Hare-Wallaby and Rabbit-Bandicoots—Travel South along the Finke and across the Missionary Plains to Hermannsburg—The Mission Station and its Influence on the Natives—Divide into Three Parties—Follow the Finke through the James Range to Palm Creek—Three Days Camp at Palm Creek—Palms and Cycads—Account of the Animal Life of Palm Creek—Restriction of Species to a Small Area as exemplified by the Mollusca—Return to Hermannsburg—Jerboa-Rats and Antechinomys Leave Hermannsburg—Modification in Form and Colour of the Foliage of Acacia salicina and Mulga—Camp in the Scrub—The Main Camel Team goes on Eastwards along the Missionary Plain to Alice Springs—A Section of the Party goes North to cross the Ranges to the Burt Plain—View from the South McDonnell Range—Camp near Paisley Bluff—A Day in Camp—Various Forms of Ant Nests—Rock Wallabies—Method of Carrying the Young in the Pouch, a Severe Handicap to Marsupials in Competition with Rodents—Brinkley Bluff—Traverse the Ranges and Camp on the Burt Plain—Strike the Telegraph Line and follow it South to Alice Springs—Mr. Watt pays a Flying Visit to the Gold and so-called Ruby Fields—A New Marsupial—The Ranges at Alice Springs—The Todd River—Conlin Lagoon—Various Forms of Phyllopods and their Habits—The so-called Barking Spider—The Sound probably due to a Bird—The Presence of a Stridulating Organ in the Spider—Leave Alice Springs and travel South along the Telegraph Line to Oodnadatta . . *Page* 107

LIST OF ILLUSTRATIONS.

PLATES.

Plate 1 is reproduced from Photographs by Mr. W. A. Horn.
Plates 2-11 are reproduced from Photographs by the Author.

Plate	1.	Camel Buggy.		
		Arrival at Water	To face page	5
Plate	2.	Camp Asleep, 5 a.m.	,, ,,	7
Plate	3.	Gibbers -		12
Plate	4.	Camels Resting.		
		Desert Oaks.*		
		Grass Trees.		
		Castle Rocks		15
Plate	5.	Lake Amadeus.		
		Crown Point		33
Plate	6.	Chambers Pillar		49
Plate	7.	Ayers Rock		85
Plate	8.	Mount Olga	,,	91
Plate	9.	Redbank Gorge	,,	101
Plate	10.	Finke Gorge	,, ,,	108
Plate	11.	Palm Creek		114

ILLUSTRATIONS IN THE TEXT.

Nest of Social Caterpillar	44
Sand tube made on a leaf of Porcupine Grass by Ants to enclose Coccidæ	70
Cycads—*Encephalartos Macdonnelli*	77
Porcupine Grass—*Triodia pungens*	85
Ayers Rock—to show weathering	86
Diagrammatic section across the country from the Burt Plains in the north to the James Range in the south	103
Mulga Trees—*Acacia aneura*	122

* By mistake this Illustration is entitled " Porcupine Grass."

Through Larapinta Land

CHAPTER I.

Introductory Remarks.

Object of the Work—Members of the Expedition—Larapinta Land—Difficulties of Travel and Nature of Camels—Departure from Adelaide and Arrival at Oodnadatta—Departure from Oodnadatta—Loading and Riding Camels—Daily Programme while on the March—The Main Sections of the Journey—The Australian Steppes.

My object in writing the following narrative is to give some idea of the nature of the country through which the Expedition passed and also of the work accomplished. To do this I have availed myself of the information contained in the various scientific reports, and take this opportunity of expressing my obligations to the various writers from whose work I have gained information of which use is made in the following pages. To my colleagues on the Expedition I am especially indebted, not only for the use which I have freely made of their writings, but for much information afforded to me during the course of the Expedition. To how great an extent I am indebted to them will easily be seen by reference to the scientific reports.

My endeavour has been, without entering into too great scientific detail, to summarise in a more or less popular form the results obtained in the various branches of science, and to convey to the reader who has not travelled in Central Australia some idea of what the country is like. By those who are acquainted with the writings of the explorers of Central Australia, such as Sturt, Stuart,

Grey, Leichardt, and Warburton, and in more recent years Giles, it will be easily realised that it is a matter of no small difficulty to render any such account otherwise than as monotonous as the country through which the traveller must pass. To their accounts was added the charm attendant upon the description of travel through untrodden country in face of almost insuperable difficulties. Though away from beaten tracks, we only traversed country previously explored, and had practically no serious difficulties to contend with. We had, however, more time to devote to an examination of the different features zoological, botanical, geological, and meteorological—of the country than was possible in the case of the original explorers, so that, in certain respects, I hope to be able to give a fuller description of a limited area of the central region than has yet been written.

The Expedition left Adelaide at the beginning of May, 1894, and three months and a half were occupied in traversing the country which it was organised for the purpose of scientifically exploring.

The members of the Expedition, in addition to Mr. W. A. Horn, who accompanied us as far as Idracowra on the Finke River, and the various branches of work allotted to them were as follows:—

Professor Ralph Tate	Geology and Botany.
Dr. E. C. Stirling	Anthropology.
Professor Baldwin Spencer	Zoology and Photography.
Mr. J. A. Watt	Geology and Mineralogy.
Mr. C. Winnecke	Surveyor and Meteorologist.

Messrs. F. W. Belt and G. A. Keartland accompanied the party as collectors and taxidermists, and there were in addition the usual camp men—a cook, two white men and two Afghans in charge of the camels, and black "boys" to serve as guides.

The object which Mr. Horn had in view in sending out the Expedition was not to explore new country, but to examine as carefully as time permitted the country in and about the McDonnell Ranges. These lie almost in the centre of Australia just to the south of the Tropic, and, roughly speaking, stretch across from east to west between long. 130° and 135°. To reach them it was necessary to traverse all the district lying between them and the northern end of Lake Eyre. All this large tract of country is drained by the Finke River and its tributaries, so that, in reality and as far as circumstances and time permitted, the

Expedition may be said to have made an examination of the great Finke Basin, which, adapting the native name of the river, may be spoken of as LARAPINTA LAND.

In judging of the results of the Expedition it is only fair to remember that some two thousand miles* had to be traversed slowly, for the most part on camel-back, and that out of a total of one hundred and twenty-five days spent in the field, less than twenty were available for actually "spelling" in camp; that is, whilst during each of more than one hundred days an odd hour or two were available for collecting, the time during which we were really free to make anything like a searching investigation was of necessity very limited indeed.

In such a district as Central Australia it is not always possible to stop just when and where you want to; waterholes during the dry season—that is, the winter months—are few and far between, and certain stages have to be made to reach them. In the scrub-covered country, it must also be remembered that travelling is often slow and tedious and from a collector's point of view a camel is the most unsatisfactory of beasts.

Perched high up between heaven and earth, you may often see, say, a lizard or an insect which you are anxious to secure, but long before you can persuade your camel to sit down the animal is far away and safely hidden. The chances are, too, that you return from a fruitless search to find that your camel, which above all things dislikes to be left behind its companions, has trotted away. Anyone who has attempted the task knows well the difficulty of persuading the beast to sit down when it does not want to do so, and will sympathise with the feelings of an unexpert rider who attempts to safely mount a camel which is anxious to be up and off after its fellows.

A camel has a peculiar way of its own of getting up, which is bad enough when done slowly; but when it is in a hurry, then you have to be very careful not to get an ugly bump or fall. The moment you are in your seat behind the hump, or perhaps before you are there, he rises with a jerk half way up on his hinder legs, throwing you forwards; before you have time to recover your balance up go the front legs half way, then it rises completely on its hind legs and finally on its front legs—a fourfold movement of a most disagreeable nature. To make it sit down, the magic word "hūsht" must be repeated until it kneels down on its front legs; then it swings backward half way down on its hind legs, then completely down on its front legs, and, lastly, completely down on its hind legs.

---

* That is, the distance traversed after leaving Oodnadatta, the head of the railway line, which itself lies more than 600 miles north of Adelaide.

Then, too, the movement of the beast when it walks or trots has a peculiar churning effect on specimens, and as it is not always possible to safely stow them away when on the march, many a one is bruised and spoilt. In walking it does not move its feet like a horse—two diagonally opposite ones at a time—but the two near or the two off feet are lifted simultaneously.

In arid country, such as we for the most part traversed, the camel certainly has great advantages; but it must be confessed that you first mount your beast without any expectation of pleasure, that you derive none whatever from your association with him, and that you part company without any regret on either side.

The bull camels will fight furiously for the possession of the cows, biting each other fiercely with their powerful canine teeth. The victor, if it does not entirely disable the vanquished one, will chase the latter away at headlong speed, utterly regardless of anything in its way; and if the fight takes place at night, as it once did with us, and the flight of the vanquished one happens to be directed through the camp, then the consequences may be very serious, as two infuriated camels running "amuck" require to be given a wide berth.

A bull camel has a remarkable habit of in some curious way forcing the air in behind the uvula and forming a bladder, which begins to come out at one side of the mouth. The beast makes a loud bubbling sound, the bladder in the meantime growing larger and larger until it is as big as its head. Then the bubbling ceases, and the bladder is gradually withdrawn.

The neck is so long that when you perhaps imagine yourself well out of harm's way, you are startled to hear a sudden snap and to find that the beast has made a savage bite at you. If angry, they will try and get you down upon the ground and endeavour to pound you with the hard callosity on their chest. Altogether, it is best to be on your guard when dealing with camels; there is no getting fond of them, and of all beasts of burden they combine in the highest degree the qualities of filthiness, viciousness, and crass stupidity.

The ordinary baggage differs, as it has been said, from the riding camel as much as a thoroughbred does from a cart horse; and of all the methods of travelling, the back-breaking swing of a rough camel is the most monotonous. A good riding camel will travel as fast as ten or even twelve miles an hour, and can keep this up for many hours during the day, but the ordinary loading ones will not cover more ground than between two and three miles an hour. They always travel in single file, and it is most difficult to get two to walk side by side, so that conversation whilst on the march is conducted under difficulties.

Horn Exped. Cent. Aust. Narrative, Plate 1.

Mr. Horn's Carriage.

Arrival at Water.

However, their powers of endurance, despite their vicious disposition, render them invaluable in dry countries such as the interior of Australia. They will feed on thorny desert plants which nothing else will eat, and can, when trained, go for days together without drinking—the longest record in Australia being, I believe, the 21 days' waterless march on the recent Elder Expedition. Such abstinence as this must, however, cause considerable suffering to the animals.

However, to return to our Expedition. Leaving Adelaide, we went by train for 600 miles to Oodnadatta, the most northern point on the southern part of the projected trans-continental railway. Mr. Winnecke had preceded us to superintend arrangements, and we found the camels camped some little distance outside the township in the midst of a dry, bare plain, close to a small muddy waterhole on the Neale Creek.

Mr. Winnecke had evidently been having a busy time. Stores of all kinds and collecting material were ready, and next morning the camel train moved out of camp and took the track northwards along the overland telegraph line towards Charlotte Waters.

We had altogether some twenty-five camels and two horses, each member of the scientific staff having his own riding camel, the remainder being loaded with various weights according to their carrying capacity, the heaviest load weighing between seven and eight hundred pounds.

Perhaps the most curious part of the whole caravan was a buggy drawn by a pair of camels. This was only taken over the first two hundred and fifty miles of our journey, when we were travelling along the track by the telegraph line as far north as Crown Point, where we were not sorry to leave it behind. Out in the bush it would have been impossible for it to have travelled, and even along the rough track, where travelling was comparatively easy, it was not exactly an unmixed blessing when rough creek-beds had to be crossed. In the illustration (Plate 1) the camels are represented as sitting down in the position in which they had just been harnessed; when standing up they naturally looked very ungainly and far too big in comparison to the size of the buggy. Though in some parts of Australia, such as the West, camels are now regularly used for this purpose, they seem to be much more fitted for carrying burdens than to serve as draught-animals.

All the camels used were the single-humped ones, and the saddles are so made that they are kept in place partly by the hump itself, partly by girths. A loading camel will carry a big box on either side and another package on the top. Everything, of course, is fastened on while the camel is sitting down, the beast frequently expressing its disgust and annoyance at the process by growling and gnashing its teeth. Unless securely fastened on, the slow but steady churning movement, which is much like a combined pitching and tossing and rolling, will soon put the packages out of place.

For the first day or two, until the weights are fairly adjusted, the loads are continually shifting and stoppages are frequent. Each camel has a hole bored through one side of its nose, and into this a wooden peg is fixed, shaped something like a little dumb-bell ; to this a string is tied, and so in a baggage train a string passes from the nose of one animal to the tail of the one next in front, for of course they walk in single file.

So long as the travelling is easy this is right enough, but in difficult country, as when, for example, a creek with steep sides has to be crossed, it is not easy to avoid a break-away. The front one of the camels coming first to a steep descent and carrying a heavy load is very apt to go down with a sudden run, which probably means that the hinder one stands still and the nose-string is broken. The nose-peg itself is not infrequently pulled out and has to be replaced, or, if the string by good fortune simply comes untied (the knot is always a loose one) from the tail of the front animal, the hinder ones will stand still, sniffing the air in a stupid, idiotic kind of way, until they are led up to the front one and the damage repaired. In difficult country this often takes place, and so travelling is slow work, and the distance traversed may not average more than two or, at most, three miles an hour during the day.

Going down a steep bank a camel will often slip down on its haunches, and going up one will climb on its knees. Often there is serious difficulty in getting them to cross a creek holding water. Mr. A. W. Howitt told me of an ingenious plan adopted by himself when he was out in charge of one of the parties despatched to search for the remnant of the ill-fated Burke and Wills Expedition. He had come to a creek full of water, and the camels steadily refused to go into it. At last a happy idea struck him ; he had one of the beasts brought up and made to sit down broadside on to the creek. He and his men ranged themselves on the land side of the animal, which was then made to get up, but whilst in the act, and at a given signal when the beast was off its balance, a united push sent it sprawling into the water, across which it then made its way.

Whilst on the march our daily programme was much the same. Usually just before sunrise we were up and dressed. Very shortly after sunrise we had breakfast. Our camp cook, Laycock, was an old hand at the work, his experience

dating back to the building of the overland telegraph line; and thanks to him, so long as we remained in the main camp we lived in comparative luxury. Breakfast—always hot and most welcome—was eaten when usually the temperature was not much above freezing point. The black boys and the Afghans brought the camels into camp, and along with them the odour of their undigested feed.

Whilst the loading of the baggage-camels took place, each of us saddled and packed our own beast. A riding saddle is so made—they are wonderfully crude and heavy structures—that you can pack your personal belongings in front of the hump, while behind is a seat for yourself in such a position that the animal can, when it desires to do so, whisk its filthy tail on to your back.

The reins of a riding camel consist of two strings, one passing round each side of the neck and attached in front to the single wooden peg inserted in one side of the nose. Owing to the fact that a hard pull is liable to at once bring out the peg, this gives the rider the minimum of control over a beast so naturally stupid as the camel. More than once, when I had stayed behind the rest to endeavour to secure some particular beast or to take a photograph, my camel started off at a quick trot to catch up the train. All that I could do was to hold on to my camera and luggage and hope that the train was not far ahead; the camel was sure to reach it safely, but there was every chance of the camera and myself being left behind. I may say that I had christened my camel the "Baron," after my distinguished friend and counsellor, the Baron von Mueller, whose name is a household word with us in Victoria, in the hope that, as the bearer of such a name, he would behave himself accordingly, but I was disappointed in him.

Once mounted, we travelled slowly on at a walking pace for perhaps ten or twelve miles, with plenty of time to observe the nature of the country, but with no or little opportunity to collect. Then came a halt in the heat of the mid-day for lunch, when collecting was made difficult by reason of the flies which settled on your face. After the halt, another march of the same length brought us at dusk to our camping place for the night. The camel train was brought into camp forming a semicircle; each camel was unloaded, and then, after being hobbled, was set free for the night to find what feed it could. The camp fire was lighted, notes were written up, specimens labelled and packed away, and then we lay down and slept in the open under the perfect clearness of the desert sky. As a general rule the nights were very cold, not infrequently the thermometer registering several degrees below freezing point; but the air was so dry that the cold was comparatively little felt, even when our water-bags were frozen solid.

This programme, repeated day after day whilst traversing country of the most desolate description, soon became very monotonous; in fact, the most striking feature of travel in Central Australia is the wearying monotony which stands out so clearly in the writings of all the explorers of the interior.

Looking back upon our journey, it appears to divide itself up naturally into certain sections—*first*, the country between Oodnadatta and a little to the north of the Charlotte Waters Station, where we struck the main Finke River and its tributaries; *second*, the country along the Finke until we reached the James Range; *third*, the Silurian ridges which form the southern part of the James Range and the George Gill and Levi Ranges; *fourth*, the desert sandhill country across Lake Amadeus to Ayers Rock and Mount Olga; and, *fifth*, the interesting and varied country in and about the northern part of the James and the McDonnell Ranges.

Speaking generally, our journey led us into three types of country. It is usual to speak of the whole interior of Australia as a Desert or Eremian country, but this name as applied to the whole area is really very misleading. It is true that over wide areas extending especially across the western half of the interior there spread out sandhills and flats covered with Mulga scrub or "Porcupine" grass which may justly be described as Desert, and across which no creeks of any size or rivers run, and where water is only to be found often at long intervals of time in isolated clay-pans or in rock holes amongst the rocky ridges which every now and then rise above the sand and break the dead level of the monotonous plains.

Such true desert country has been repeatedly described in the writings of many of the Australian explorers—Grey, Forest, Warburton, etc.—and such country we passed across in the journey from the George Gill Range to Ayers Rock and Mount Olga.

But, in addition to this true desert, there is a vast tract of country comprising the great Lake Eyre Basin, stretching from this eastwards and northwards into the interior of New South Wales and Queensland and up to and beyond the McDonnell Ranges, across which run such intermittent streams as the Cooper, the Warburton, the Macumba, the Finke, and the Todd, dry for the greater part of the year, but every now and then at varying intervals of time swollen with heavy floods which spread out over wide tracts, and for a time transform the whole country into a land covered with a luxuriant growth of vegetation. To this part of the continent the name of the **AUSTRALIAN STEPPES** may be suitably applied.

Starting from Lake Eyre, and travelling northwards towards the centre of the continent the traveller passes across a tract some four or five hundred miles in width which may again be divided into two districts, which may be called respectively the **LOWER STEPPES** and the **HIGHER STEPPES**.

The **LOWER STEPPES** extend over the area occupied by the great Cretaceous formation with its alternating stony or gibber plains, loamy flats, and low-lying terraced hills capped with Desert Sandstone. At Lake Eyre the land is thirty nine feet below sea level, and gradually rises to a height of one thousand feet at its northern limit.

The **HIGHER STEPPES** are characterised by high ridges of Ordovician and Pre-Cambrian rocks which stretch across the centre of the continent from east to west for some four hundred miles. The average elevation of these Higher Steppes may be taken as about two thousand feet, and above them the higher peaks of the ridges rise for some two thousand five hundred feet more.

Both the Lower and the Higher Steppes, as already said, are traversed by creeks and rivers which are absent in the true Desert Country. In the following account the Lower Steppes are described in the chapters dealing with the country between Oodnadatta on the south and the James Range on the north, and the Higher Steppes in the chapters dealing with the James, George Gill, and McDonnell Ranges, and the Desert Region in the chapters describing the journey from the George Gill Range across Lake Amadeus to Ayers Rock and Mount Olga.

The remarks of Brehm* are exactly applicable to the centre of Australia. He says, "In order to understand the steppe lands it is necessary to give a rapid sketch of their seasons. For every country reflects its dominant climate, and the general aspect of a region is in great part an expression of the conflicting forces of its seasons, apart from which it cannot be understood."

Now the climate of Central Australia is one which reveals an alternation of short rainy seasons with intervening periods of drought. The rainy season is short, the dry season long, and not only this but, whilst the rain season is always short the dry season may be abnormally prolonged. There is no regular succession of spring, summer, autumn and winter, but simply a hot and a relatively cold season, that is a summer and a winter with a longer or shorter interval during the former when the rainfall takes place.

---

* "From North Pole to the Equator." English translation by Margaret M. Thomson, p. 169

Further still, the land is one where almost perpetual sunshine reigns; week after week, often month after month the sun shines brightly all day long in an almost cloudless sky. In the summer the heat is intense, but in the winter months from May to September, whilst the days are very hot the nights are bitterly cold—the temperature often falling many degrees below freezing point.

To this irregular alternation of seasons, and to a great diurnal variation in temperature every animal and plant must become adapted if it is to survive. Hence it is that so many of the plants are those which have special provision to prevent rapid evaporation of moisture—such as the spiny Acacias and grasses, the wiry Casuarinas, the hairy-leaved Atriplex or salt-bush, and the succulent Claytonia and Portulaca which have thick cuticles.

In addition to the special modification of the adult plants, the seeds require to be of such a nature that they can both withstand the influence of long exposure and at the same time germinate rapidly directly the conditions become favourable. Anyone who has seen the inland loam flats and even the stony gibber plains, bare and desolate before the rains and green and luxuriant a few days afterwards, will realize the phenomenal rate of germination and early growth possessed by many of the steppe plants.

Amongst animals we find the kangaroo and the dingo, which can travel long distances with ease, or else, like the native blacks, can subsist, if need be, on the dew which in early morning condenses on the grass, smaller marsupials which can feed upon the ants or dried up vegetation, frogs and mollusca which remain hidden in the damper ground beneath the hard-baked surface, and crustacea such as Apus and Estherias, the eggs of which will not develope unless the water in which they have been deposited dries up.

CHAPTER II.

The Lower Steppes.

From Oodnadatta to Charlotte Waters and the Finke River.

Lake Eyre in the Dry and Wet Seasons—Gibber Plains—Origin of the Gibbers—Loamy Plains—The Valley of the Macumba River—Water Holes—Chestnut-eared Finches—The Prickly Seed Cases of Tribulus and Bassia—Succulent Plants, Claytonia and Portulaca—Remarks on Spinous and Succulent Development of Plants Both forms of growth are probably adaptations to climatic environment and not in the first instance developed as protection against animals—The most spiny and the most succulent plants are found in the arid regions—The Stevenson River—Contents of a Water Hole during the Dry Season—Tenacity of life of Bithinia australis—Dalhousie Station and Mound Springs—Red Mulga—Gibber Plains at Sunset—Clay Pans; contrast between them in the Dry and Wet Seasons—The Fauna of a Clay Pan—Amphibia, Crustacea, Mollusca—Colour Changes of Frogs—Habits of Apus—Fresh Water Crab—Water Holding and Burrowing Frog—The Adminga Creek—Giddea Scrub—Charlotte Waters Telegraph Station—A Second Visit to Charlotte Waters in Summer Time—Flies and Mosquitoes—Succession of Forms of Life—The Colouration of Lizards—Sexual Differences—Brilliant Colouration, the accompaniment of a general state of activity and only indirectly associated with that of the environment—Susceptibility to Heat of Lizards—Tiliqua occipitalis killed by Heat of Sand—Departure from Charlotte Waters—Change in Nature of the Country—Ant Lions—Mount Daniel—Camp at the Goyder River—Habits of Physignathus longirostris—Messrs. Watt and Winnecke start off to follow up the Goyder and Lilla Creeks—The Main Party goes on to Crown Point—View of the Finke Valley.

On its way north the railway line now passes close to the western border of South Lake Eyre, and at this point is actually some three or four feet below the sea level. As a general rule the Lake is for the most part, as it was when we passed it going and returning, a white sheet of salt. Into it drain the more important rivers of the interior—on the west the Barcoo and Warburton, on the north the Neale and the Macumba, whilst in times of heavy rain amongst the ranges in the centre the flood waters of the big Finke itself probably help to swell those of the Macumba.

It is only after very heavy rains that these rivers run, and then the Lake bed is filled with water, as it was when I passed by it in January, 1895. Then the stony plains around were green with grass, and the waves, blown by a heavy wind, were breaking in spray against the small cliffs bounding the shore. The evaporation is, however, so great that only a comparatively short time passes before all is once more dry and parched.

From Oodnadatta our course lay across a gradually rising and somewhat undulating country with low-lying flat-topped hills and upland plains covered with " gibbers."* These gibber plains, a characteristic view of which is shown in the accompanying illustration (Plate 2), are the most striking feature of this part,

---

* The name is derived from a Queensland aboriginal word "gibber," which means a stone.

and in all probability are identical with the "stony desert" of Sturt's description of the interior.

Stretching away to the horizon on every side is a level plain covered with a layer of purple-brown stones, varying in size from an inch to perhaps a foot in diameter, all made smooth by the constant wearing away of wind-borne sand grains. Amongst them in the dry season are here and there a few small tussocks of yellow grass; small lizards dart about, and innumerable grasshoppers rise up from your feet and fly for a short distance. There is no water and no shelter; perhaps a line or two of thin mulga trees far away will mark the course of a dry stream which meanders about for a short distance as it comes down from some low lying hill, only to be soon lost upon the plain. Except within a short time after rain it is useless to look along its bed for water holes. The surface is dry and cracked, and where the water stood longest are curled flakes of a glistening clayey nature.

Nothing could be more desolate than a gibber plain when everything is bare and dry, and the outline of the distant horizon is indistinct with the waves of heated air.

Throughout all this district the low flat-topped desert sandstone hills indicate the original level of the land. All these hills have a thin capping of hard chalcedonized sandstone; when once this is broken up the softer underlying rock is rapidly disintegrated, and the sand particles into which it breaks up are partly carried away in flood time, and partly blown away by heavy winds.[*] The harder chalcedonized material gradually breaks up into blocks of various sizes, and these become polished and rounded by the wind-blown sand grains, while a thin coating of oxide of iron gives them a red brown and curiously polished appearance. As the sand is gradually removed the polished stones come to form a layer spread over the flat surface of the plains, the stones of which are so close to one another and so regularly arranged that at times they look almost like a tessellated pavement. In passing from the plains up the sides of the hills the gibbers can be seen in all stages of formation, from the small, smooth and flattened pebble on the plain to the big, irregularly shaped mass which has just tumbled off from the exposed surface of the thin desert sandstone capping of the hill.

These stony gibber plains merge constantly into loamy plains covered with poor scrub, but on which the gibbers are wanting. Perhaps, as suggested in the section dealing with Geology, these loamy plains occupy areas on which the Upper

---

[*] Strong south-east winds during the winter months.

Cretaceous rocks were not capped with the hard chalcedonized Desert Sandstone, and where, therefore, no gibbers have been formed.

Shortly after leaving Oodnadatta the track passes away from the telegraph line, leaving the latter some miles to the west. We crossed two or three smaller creeks, such as the Opossum and Storm Creek along which a few water-holes still remained, and after three or four days came into the broad valley of the Macumba, which during the winter months simply forms a succession of dry sandy beds running parallel to one another with muddy water-holes here and there, which after a few months of drought dry up completely. The approach to a water-hole can always be told, not only by the greener patches of scrub and trees immediately surrounding the water, but by the twittering of innumerable chestnut-eared finches (*Tæniopygia castanotis*). The twittering of these pretty little birds may always be taken as an indication that water is not far away: from the side of a water-hole flocks rise as you approach, and their little grass nests are very common, as many as nine being seen on one occasion on one small shrub. They fall an easy prey to such birds as the falcons, which will swoop down upon a flock and usually carry off a little finch each time. Judging by their numbers they must be prolific breeders.

There is not, however, much life as a general rule about these water-holes, and a yard away from them everything is as dry and parched as possible. In the dry season the only moist place in Central Australia is actually in a water-hole.

The lines of the water-courses are marked by belts of gum trees and acacias —*Eucalyptus rostrata*, the river gum; *Eucalyptus microtheca*, the swamp gum; *Acacia aneura*, the mulga; *Acacia cyperophylla*, the red mulga, a very local tree extending across a narrow belt of country from east to west, a little way to the north of the old Macumba Station, and the stinking acacia, *A. homalophylla*. On the loamy flats, and even gibber plains, the most noticeable plant is *Salsola kali*, popularly known as the Rolly-polly. It is, when mature, one of the characteristic prickly plants of the Lower Steppes, and forms great spherical masses perhaps a yard or more in diameter. It is a constant feature of the Cretaceous area, and gradually disappeared as we passed northwards into the Silurian district.

The thin, poor scrub is made up largely of Cassias, Eremophilas, Hakens, and Grevilleas, all thinly scattered about, and with hard, spiny or coriaceous leaves. Now and again, especially on the upland stony plains, were patches of salt-bush (*Atriplex rhagodioides*), the foliage of which has the characteristic and well-known blue-grey tint, caused by the presence of a "mealy" secretion on the leaves, which is probably of service in checking too rapid evaporation. The ground is not like that which one is accustomed to in moister parts; tussocks of grass, such as *Spinifex paradoxus*, are scattered about, with little plants of the red stemmed and poisonous *Euphorbia Drummondii*, or of one or two species of Ptilotus (*P. exaltatus* and *incanus*), but they are not crowded together, and you can count the separate plants. It was not at all unusual to see a small patch of ground occupied entirely by a colony of one species of a plant such as Ptilotus. Along by the river flats the clusters of red fruit of the Darling or Murray Lily* were frequently seen, whilst in the wet season its white flowers are a striking feature along the Stevenson Valley.

When once they have grown to a certain size, none of the plants growing on the Cretaceous table-lands and along the flats bordering the creeks have to compete with one another for space on which to grow. The question of which are to survive and which are to die is settled in the main at a very early stage, when they are seedlings. Directly after the rains have fallen the ground is thickly covered with the bright green of endless seedlings, but it is only those which can reach a certain size and stage of development before the dry season fairly sets in which have any chance of surviving, and at a very early time the weaklings die off and the stronger ones are left to grow up with no competition as between plant and plant, but with a hard struggle against climatic conditions.

When we passed through, in the dry season, one of the commonest plants on the ground was a creeping species of Tribulus (*T. terrestris*); its large yellow flower is pretty enough, but its dried and prickly seed cases are more than irritating when you try to camp amongst them, and they seemed to be with us always.

Quite as irritating, though happily not quite so plentiful as the Tribulus, are various species of Bassia. The seed-cases of these have a pretty downy centre, perhaps half an inch in diameter, but around this are a number of very stiff, sharp-pointed spikes projecting through the soft down. What with these and other prickly seeds our camping place was often a bed of thorns, and after selecting a spot, a usual preliminary to opening out our rugs was to sweep the ground with an impromptu broom of Cassia branches.

Whilst many plants in the arid and desert regions are protected against too rapid evaporation by having their leaves or leaf-stalks transformed into thin, switch-like structures, others go to the opposite extreme and become thick-leaved and succulent. The most common of the latter in the district through

* This is an Amaryllid plant *Crinum flaccidum*.

Polo Cross Arcadia Junior.

Castle Rocks.

Camels Resting.

Grass Trees.

which we travelled are species of Portulaca, popularly known as munyeru, and various species of Claytonia. These grow in little clumps, lying low down upon the ground, and remain soft and juicy when everything else is dry and withered.

There can, I think, be little doubt but that this switch-like structure of leaves and leaf-stalks, together with, in the case of the desert oak (*Casuarina Descaineana*), the loss of leaves, and the substitution for them of little still green twigs, and also, in other plants, the development of hard, thorny processes around the seed-cases, is simply due to an adaptation to climatic influences, and has, in the case of the Central Australian plants, very little, if indeed anything whatever, to do with protection against animals.

In the first place, there are comparatively few animals to feed upon them; kangaroos and wallabies and other plant-eating marsupials do not exist in anything like sufficient numbers to keep the plants down; and then those which are succulent and edible, such as the munyeru and Claytonias, and in no way protected against animals, so far as can be seen and judging from the way in which they eat them, thrive just as well as the spiked and thorny plants.

What appears to be most probably the case is, not that the prickly growth is brought about in any way as a protection against predatory animals, but that it and the succulent development as well, are adaptations to suit climatic environment. If animals, so to speak, want to feed upon these climate-proof plants, then they must become fitted to do so. None of these Central Australian plants, which are as spiny as they can well be, are in the least thereby protected against such an animal as the camel, which will, with relish, munch away at the most thorny Acacia (*Acacia farnesiana*, for example) just as readily as it will feed upon the juicy Claytonia.

It is at all events worth noticing that it is just in the hot, more or less arid and desert parts where animals are least numerous, that both the spiny and the especially succulent plants are best developed, and it seems reasonable to connect this with their climatic rather than with their animal environment.

After crossing the Macumba our course lay northwards along the valley of the Stevenson, the Macumba River being formed by the union of the Stevenson coming down from the north, and the Alberga which runs in from the east, having its principal source probably in the Musgrave Range. As usual the river was simply a sandy bed with a few water-holes at intervals.

Our camp for the night was pitched when possible by the side of a water-hole. These are all very much like one another. A patch of green scrub lines their banks, and in the water will be found a fair number of molluscs, such as species of Bulinus, Bithinia, and the common mussel, of which the blacks are very fond, one or two species of Estheria, and water beetles in abundance, with probably a frog or two. On their muddy margins fresh water crabs will sidle away towards their holes in the banks. Plenty of little chestnut-eared finches will be flying about amongst the shrubs, and perhaps a pair of graceful dotterels (*Egialitis nigrifrons*) may be seen running about in search of aquatic insects. These are all the animals that will be found in and about such water-holes as exist for some time during the dry months. On our way back, some four months later, almost all the water-holes in this district were dried up, but buried in the dry clayey mud forming their beds were clusters of operculate molluscs* and numbers of water beetles alive. The crabs had apparently all retreated into their burrows, but the Estherias were all dead and their empty carapaces strewn on the surface.

After a day's travel beyond the Macumba we turned off slightly to the east so as to pass the outlying station of Dalhousie. If possible the country was more desolate than ever—long upland, gibber plains with bare flat-topped Desert Sandstone hills. Across this part are scattered the well known mound springs. These mounds are often of considerable diameter, perhaps upwards of 50 feet in height with a pool of often warm and sometimes even hot water on their summits. The water is more or less impregnated with mineral matter brought up from below, and it is the deposition of this which has gradually formed the mound as the water evaporates and the sinter or travertine is left behind. At Dalhousie the mound around the spring was black with decaying vegetable matter, for the pool was surrounded with a growth of rushes. Over the side of the mound the water trickles down, but the channel thus formed only extends for a short distance as the evaporation is too great and the water supply too small to form anything like a long stream.

These mounds of sinter or travertine, capped with green vegetation, form a striking feature in the otherwise dry and parched-up country in which they are found.

A little to the north of Dalhousie we crossed a narrow belt of country characterised by the growth along the creek sides of red mulga. This is an Acacia (*A. cyperophylla*) reaching perhaps a height of twenty feet, the bark of which, alone amongst Acacias, is deciduous and peels off, forming little deep-red coloured

---

* Some specimens of *Bithinia australis* which I took from the bed of a dried up water-hole and put into a tin match box were alive fifteen months after my return to Melbourne, having been shut up in the box in my laboratory all the time.

flakes. It is evidently very local in its distribution, and we met with it nowhere else except in this district.

Travelling over this country during the daytime, with its dried up creeks and stony gibber plains, there is little which looks picturesque; but at sundown the scene becomes quite changed, and it is hard to believe that the picturesque appearance is due simply to atmospheric conditions.

In the desolate gibber country near the Macumba the effect was really beautiful. Away to the east the land rose to flat-topped, terraced ranges. In the foreground were white-blue salt-bushes, with pale, light blue patches of low herbage and still lighter tufts of grass amongst them, standing out in strong contrast to the purple-brown gibbers. The country was crossed by dark lines of mulga, marking the creek beds and streaking away up to the hills, which stood out sharply against a cold steel-blue sky, melting above into salmon-pink and this into deep ultra marine. In the west was a rich after-glow, against which the stony plains and hills looked dark purple, with the mulga branches standing out sharp and thin against the sky.

The colours of the Central Australian landscape at sunrise and sunset are just those which at morning and evening light up the barren ranges of Arabia—everything is soft and brilliant, but very thin.

One of the most striking features of the central area, and especially amongst the loamy plains and sandhills, is the number of clay-pans. These are shallow depressions with no outlet, and varying in length from a few yards to half a mile, where the surface is covered with a thin layer of clayey material, which seems to prevent the water from sinking as rapidly as it does in other parts.

For the greater part of the year they are perfectly dry with a thin surface film broken up into curled glistening flakes or, where the clayey mud is thicker, fissures perhaps a foot in depth run down between roughly hexagonal masses of hardened earth, which on their surface bear the imprints of the animals. Emus or Kangaroos—which crossed them while they were still moist, in search of the last remnants of water.

As we passed by these in the dry season everything was parched and silent, with no sign of animal life. The dead shells of molluscs, the carapaces of Estherias, and the foot marks of frogs showed that they had once contained an abundance of animal life. Their margins were bordered by withered shrubs of Chenopodium, by tussocks of yellow dried up grass and often by the dried leaves and hard wooden seeds of the Nardoo plant.

A few months later as I passed through the same district, soon after a heavy fall of rain, the whole scene was changed. Everything was green and bright and teeming with life. All the trees and shrubs had put on a fresh growth of leaves, the ground was covered with a rich crop of grass amongst which were acres of clumps of white flowering Amaryllids (*Crinum flaccidum*), the creeks and clay-pans were filled with water, birds of various kinds—wood duck, teal, water hens, plovers, and many others were to be counted by the score. These birds appear with the rain, and then as the water-holes dry up disappear as quickly and mysteriously as they have come.

The clay-pans were now filled with a distinct and abundant fauna of their own. Day and night they were alive with the croaking of frogs; Estherias and water beetles were darting up and down; hundreds of Apus were swimming about or else scooping out the sand on the margins of the water-holes and so making little holes in which they simply lie and die as the water rapidly dries up.

The whole change from sterility to exuberant life had taken place as if by magic within the space of only a few days.

It is worth while noticing in more detail the water-hole and clay-pan fauna of the Central area, for probably it is very similar in its nature over the whole of the interior, and it consists of representatives of three groups of animals which have, each in its own way, become especially adapted to the climate of the steppes and desert with their long seasons of drought and short intervals of rain.

These three groups are the Amphibia, Arthropoda, and Mollusca*.

To begin with the Amphibia. Standing by a water-hole or clay-pan though you can hear the frogs croaking all around you cannot so easily see them. The surface of the water is flecked with the long stalked floating leaves of the Nardoo plant (*Marsilea quadrifolia*) which are fully grown, while the permanent short stalked leaves around the base are as yet only beginning to develope and are covered with water.

If you disturb the water you will see a number of little green patches, which you have probably taken for Nardoo leaves, suddenly disappear. These are the heads of one or two kinds of frogs (either *Chiroleptes platycephalus* or *Heleioporus*

---

* This refers to the water-holes and clay-pans in the desert and stony table-land country which are of temporary nature and not to the fewer deeper and more permanent rock-pools amongst the Ranges. I have purposely omitted Fish because they do not form part of the permanent fauna of these water holes and clay-pans, being only washed down into them during flood times from the permanent pools amongst the Ranges, or perhaps carried about in the form of eggs attached to the feet and feathers of birds.

pictus) and you are all the more surprised because, if you have only seen them before in the dry season, you were not at all prepared for such a transformation in colour. Then they were a dull, dirty yellow like the water and the dried up banks and vegetation, now they are yellow and orange and green like the water which is thick with yellow sand and mud particles, and dotted with bright green Nardoo leaves. Both these frogs are a fair size, but, in addition to them, there will be found a good many little grey and brown Hylas (*H. rubella*) sometimes brightened with yellow patches, but, on the whole, dull coloured in both the wet and dry season. They will be found hopping about on the banks and hiding in damp places under stones and, in addition, hundreds of tadpoles will be seen which have developed with great rapidity from eggs deposited since the rainfall.

I am much indebted to Mr. Alexander Sutherland who has been good enough to inform me of some of his interesting results recently arrived at in the matter of the varying rate of development of frog eggs at different temperatures from which we can form some idea of how rapidly the eggs develop in a Central Australian water-hole.

In a letter which Mr. Sutherland has kindly allowed me to reproduce he quotes the following results of experiments on batches of eggs of *Hyla aurea* consisting of thirteen in each.

Experiment A.		Experiment B.	
Average Temp.	Time.	Average Temp.	Time.
26·6°	39 hours	26°	48 hours
25·2°	50 ,,	24·3°	52 ,,
23·9°	59 ,,	23°	56 ,,
22·1°	69 ,,	22·2°	65 ,,
21·5°	73 ,,	21·7°	67 ,,

In another experiment the average temperature was 30·8° and the time occupied in hatching out was 34 hours; in another the average temperature was 30·7° and the time 31 hours; and in another the average temperature was 28·7° and the time 37 hours.

Mr. Sutherland adds "thus if these eggs are to hatch out in three days the temperature must be only between 21° and 22°. Now, in my present turtle egg hatching experiments, water kept without artificial heat in a cellar shows a range of only 18·5° to 21° after four days of observation taken day and night at intervals of three hours. I should not be in the least surprised if the ponds in Central Australia reached 25° as a tolerable average through the summer months in which case two days would be enough to hatch out the eggs. If the hot day lasted sixteen hours, and heated a pond to 27°, while the night in which the water cooled to 18° lasted eight hours, then an easy calculation would show that the time should be about fifty-five hours. Probably it is an essential to the reproduction of these creatures that they should spawn in hot weather and so secure the advantage of a two days period of incubation."

I have not yet had the opportunity of testing the rapidity with which the frogs' eggs develop in the clay-pans and water-holes of Central Australia, but as the rains fall during the hottest part of the year, when even at night-time the temperature remains high, there can be little doubt that the temperature of the water is exceedingly favourable to a rapid development, and there is no doubt whatever that this rapid development does take place; in fact, if the animal is to have any chance of surviving it must do so.

Amongst the Arthropoda the most striking form is Apus (*A. australiensis*), which is often seen coming to the surface, where it swims about on its back, its red appendages rendering it easily seen from above, whilst from beneath its yellow carapace may perhaps serve at once to hide and to protect it from its enemies, the voracious water-beetles, which are darting up and down. Various species of bivalved Crustaceans, some three-quarters of an inch in length, swim about. One form, *Estheria packardi*, is present in great numbers and persists long after the other forms have disappeared from the water and are represented only by their empty carapaces. This and some of the others have red blood, but the larger forms, which are much rarer (belonging to a new genus, Limnadopsis), have quite colourless blood.

All these Crustacea for some reason seem to prefer muddy water. From the Macumba River, during the summer time, when it was in flood and the water was muddy, I secured specimens of all of them, but searching in the same water-holes two or three weeks later, when the water was clear, there was not one to be found though they were still alive in the muddy clay-pans close by.

The contrast between the way in which Apus and the Estherias swim is very marked, the former on its back with the feet uppermost and the latter with the feet lowermost. The difference is probably associated with the fact that the two halves of the Estheria carapace can be completely closed over the animal's body for protection, whilst such closure cannot take place in the case of Apus, whose soft and blood-red appendages are very prominent and would be constantly seized upon by the voracious water-beetles if it swam on the surface with its back uppermost. On one occasion, as noted in the zoological reports, I came across an Apus struggling violently and on taking it out of the water found no fewer than three water beetles tearing its soft appendages out of which the blood was oozing.

In the water-holes along the creeks, but not in the clay-pans, the banks are thick with the holes burrowed out by the fresh water crab (*Telphusa transversa*) the distribution of which so far as at present recorded is a curious one as it has only been described from the central region and the very north of Queensland at Cape York and Thursday Island. In all probability it is widely dispersed over the interior of Queensland and New South Wales, though the contrast in its surroundings at Cape York and Charlotte Waters, for example, is as marked as it can well be.

Amongst the Mollusca forms belonging to the genera Bulinus and Bithinia will be found attached to any bit of stick or weed, and the fresh water mussel (*Unio stuarti*) is sometimes present in abundance buried in the muddy banks of the creeks, though neither it nor the crab are found in the proper clay-pans—that is in the shallow depressions not in the course of a river bed.

In addition to these animals there are often seen little light brown jelly like masses, which when alive I took to be fresh water Sponges, but which on further examination turn out to be colonies of Rotifers (*Lacinularia* sp.) some of the colonies reaching a length of an inch and a half, and in addition to these a branching Polyzoon is often found attached to stones and sticks.

Sooner or later the clay-pans and water-holes dry up, and to all appearance animal life has completely died out. In the case of the Estherias, Apus, Rotifers, and Polyzoa the animals have all perished, but their eggs remain and can be blown about from one place to another by the strong winds which often prevail throughout the dry months, and they are ready to develope as soon as ever the water-holes are again filled. In the case of the other members of the clay-pan fauna it is quite different, for if you know where to look for them you will be able to find them hidden away safely æstivating. They have one and all gone down into the mud while it was soft and in this which becomes so hard that you can only break it away bit by bit they lie imprisoned until released by the next heavy rains. Probably many of them perish if the drought be of exceptional length. The most interesting animal is the Burrowing or Water-holding Frog (*Chiroleptes platycephalus*). As the pools begin to dry up it fills itself out with water, which in some way passes through the walls of the alimentary canal filling up the body cavity and swelling the animal out until it looks like a small orange. In this condition it occupies a cavity just big enough for the body and simply goes to sleep.

When, with the aid of a native, we cut it out of its hiding place the animal at first remained perfectly still with its lower eyelid completely drawn over the eye giving it the appearance of being blind, which indeed the blacks assured us that it was. It is said that a black fellow when travelling over such country as this where in times of drought there is not a drop of water visible will use these frogs as a water supply. A native will tell you at once where to dig for a frog, being guided by faint tracks often indistinguishable to the unpractised eye of the white man. He will also obtain water from the roots of certain mallee gums and other trees, such as the Hakeas and Casuarinas. A white man may search in vain for such water supplies but a black fellow will know by instinct where to find them.

The snails protect themselves if they have no natural operculum by filling up the mouth of the shell with a pellet of hard earth, and in the case of one species, *Isidorella (Bulinus) newcombi*, one of the most abundant of the fresh water snails, which I dug out from the earth at the base of a gum tree above the water level of the quickly evaporating pool, I found that the earth seemed to have been specially prepared and finely ground down by passing through the alimentary canal. The plug thus formed had the colour and consistency of hard chocolate and was very different in appearance from the surrounding earth.

As to the water beetle (*Hydrophilus albipes*) this seems to be the hardiest animal of all, it simply goes down into the earth and there it remains in a crack making no special burrow or provision for itself. How long it can remain alive in this state is not known, but the blacks assured me that it would come out alive when the rains came.

This brief account will serve to give an outline of the natural history of a typical Australian clay-pan and water-hole, the animals living in which must adapt themselves to alternate conditions of drought and flood often recurring at irregular intervals of time.

All along our course from Oodnadatta to Charlotte Waters the country was in a miserable condition with water-holes rapidly drying up, whilst the dead and dried up carcasses of cattle which had crept under the shelter of a mulga tree to die were often seen, and showed how severe had been the drought before the last rainfall.

Our seventh night out from Oodnadatta we camped beside a water-pool on the Adminga Creek, which was bordered for the main part by a belt of the stinking

Acacia or Gidgea (*A. homalophylla*). When the branches are freshly cut it well deserves the former name as they have a most objectionable smell. Instead of having stiff, thin stems like the Mulga it has curved and twisted ones and the younger trees form more or less thick bushes. The foliage is a light ashen green and most depressing in appearance, especially when the hot sun shines upon it.

The next morning after photographing the group represented in the illustration (Plate 1), we crossed the bed of the creek, and after traversing one or two flats and very rough stony country, the track rose until close to the base of Mount Frank on the west, we once more cut the telegraph line. For miles ahead it could be seen streaking away like a thin line to the horizon on which we could just distinguish through the waves of heated air the outline of the telegraph station. Small low-lying hills seen across these upland plains during the heat of the day become transformed into high ranges and mere shrubs become forest trees reflected in the waters of the mirage lakes.

Late in the afternoon we reached the station where we were welcomed by Mr. Byrne and, after communicating with friends in Adelaide, Melbourne and Sydney, went on to camp beside the Coglin Creek about a mile to the north of the station.

We had been travelling slowly and it had taken us some eight days to traverse the one hundred and fifty miles which we had so far covered, but to most of us it was our first introduction to the interior of Australia and our time had not been wasted. Though this part lying along the telegraph line is the Central Australian highway (Plate 2) still everything we saw—scenery, plants and animals—was more or less novel to us and already a good many new forms had been collected, facts noted and we had begun our work in earnest.

Some nine months later I traversed the same district after rains had fallen for the purpose of completing work which I had not been able to finish during the Expedition and of securing certain forms, such as the Apus for example, which can only be obtained after rain. Charlotte Waters then became well known to me and I take this opportunity of expressing my thanks to my friend Mr. P. M. Byrne not only for the warm hospitality extended to me but for his most valuable co-operation in the work of collecting, especially in regard to the Mammals and Lizards of the Charlotte Waters district. Many of the more valuable and rare species have been secured since the return of the Expedition, for the simple reason that to secure them needs different climatic conditions to those which we encountered, and the opportunity of obtaining these I owe entirely to Mr. Byrne

On the occasion of my second visit to Charlotte Waters it was almost impossible to believe that I was passing over the same parched and dried up country which we had previously traversed. The contrast with respect to the vegetation and the water-holes and clay-pans has been already alluded to but in every respect the change was most striking.

Being summer time, the climate was rather trying. Even in winter during the hot days the flies are rather annoying, but in summer they are simply exasperating and all day long you must shield your eyes, ears and nose if you are to have anything like comfort. The only way in which I found it at all possible to make any observations or to collect was by tying my head into a muslin bag and putting up with the irritation on the hands. Long before the buzzing of the flies ceases in the evening the mosquitoes are humming around in myriads, and when camped out at night the only chance of sleep, unless by good luck a wind was blowing, was to lie in a little coffin-shaped tent of cheese-cloth. If the wind blew, then there were certainly fewer flies, but everything you had—clothes, food and collecting material—was penetrated by fine sand-grains. It was often in the summer time an alternative as to whether our meals would consist of bread, meat and flies, or bread, meat and sand. The blacks, whose greasy skin has a great attraction for the flies, do not seem to mind them and often you will see their eyes covered with the insects which they do not even take the trouble to brush off.

There is, however, one pest which is far less troublesome immediately after the wet than during the dry season, and that is the ants, at least this is so in the country through which I travelled. On our Expedition little black ants were wandering about everywhere, on my second visit scarcely one was to be seen. As I went up, the ground was alive with countless numbers of caterpillars of various sizes crawling about in all directions and affording a plenteous food supply not only to frogs and lizards whose bodies were swollen out with them, but also to the blacks. On the return journey not a trace of them was to be seen, but their place was taken by a particular kind of small brown grasshopper, the larger forms of which insect (such as *Trigoniza maculatus*), so plentiful during the dry months, were not now to be found. Probably these small ones in their turn would disappear and give place to something else.

Lizards abounded and were all full of eggs and not only this but just like the frogs they were, as compared with those previously obtained, in their brightest colours.

In some forms, such as *Amphibolurus pictus*, a coloured drawing of which accompanies the article of Messrs. Lucas and Frost in the Zoological section of the work, the males could now be always readily distinguished by their more strongly marked colouration from the females. In *A. maculatus* the difference is still more striking, the jet-black patches on the under surface of the male are entirely wanting in the female, and the two sexes can be distinguished at a glance.

This often really brilliant colouration has apparently nothing to do with the colour of their environment, indeed to human eyes it renders them more easily seen, and this at a time when their enemies the birds are especially abundant.

Adaptation to their environment for protective purposes is evidently by no means the principal determining factor in the colouration of these lizards. This brighter colouration which is strongly marked in both sexes but especially, as usual, in the male is to be associated with the peculiar activity of the chemical processes taking place in the skin as in all parts of the animal. In the dry season food is scarce and the animals become lethargic and dull coloured, in the rain season food is abundant, every animal is at work gorging itself, all its activities are at the highest pitch and intimately associated with the sum total of its activities and the necessary great increase of chemical activity in every organ and part of the body is the development of brightly coloured pigments. That these, as in the case sometimes of the frogs, may fit in with the colours of the environment and so, perhaps, to a certain extent, serve for protection is a secondary matter. Anyone who has collected such animals as *Amphibolurus pictus* will have brought home to him the fact that brilliant colouration is often the accompaniment of a general state of activity, and that it has, at all events in many cases, nothing whatever to do with that of the surroundings.

Though in the dry season a general yellowish colour is characteristic of many form (such as the species of Amphibolurus alluded to) which are found on the sand hills and stony and loamy plains, still there is really no difficulty, so far as human eyes are concerned, in seeing a lizard, and, in the breeding season, they become brightly tinted with colours such as blue which does not exist in their environment.

Forms such as *Gehyra variegata* and *Heteronota byucci*, which are often beautifully coloured, habitually, at all events in the day time, stay under logs and stones and are never seen in the open. One form—a new one—which we found (*Varanus gilleni*) climbs the trunks of desert oaks and gum trees, and with its purplish-grey tinge may perhaps secure a certain amount of concealment; but if you are on the look out for them it is really very rarely that you find yourself deceived by colour markings, and in some cases, as, for example, the bright red tail of *Ablepharus ruficaudatus*, the colour is a decided help.

It is, of course, possible that the red tail may be easily seen and pounced upon by an enemy, who secures the tail but loses the body, but it is, on the other hand, difficult to understand what particular advantage the possession of a conspicuous part of the body is as compared with the advantage to be gained from a general inconspicuous colouration of the whole body.

Then too, as in the case especially of *Egernia whitii*, there is a very great range in colour amongst specimens found in the same district as they may vary from a dull yellow-brown with strong black markings to a bright brick-red with faint dark markings—a variation which has nothing to do with their surroundings.

In the case of some of the beetles, as, for example, many of the grey Curculios, which lie quiet in the cracks of bark, the colour of which they exactly assume, it is certainly not an easy matter to always determine at first sight whether you see a beetle or a bit of bark, but then, supposing these fall a prey to such an animal as a lizard, the latter climbing a tree trunk, or a bird doing the same, will probably be guided quite as much by the sense of smell as by that of sight.

It would not, of course, be a difficult matter, so far as these Central Australian animals are concerned, to gather a series during the dry season and place them amongst sand and stones and withered herbage as an illustration of protective colouration, but then it would be only right and equally instructive to take identically the same series during the wet season and place them amongst their surroundings as an instance of the general absence of any special protective colouration.

I have already pointed out however, that the frog, *Chiroleptes platycephalus*, does, without doubt, at different seasons assume a colouration which is in general accord with its surroundings; but whilst this must be admitted to be the case, there are other considerations which must be taken into account. At first sight the head of a Chiroleptes looks very much like a floating Nardoo leaf, but a very small amount of experience enables you to distinguish between the two, and, in addition to this, the frogs are in much more danger from their enemies on land than in water. Now, the slightest rustle near water makes them disappear at once, and on land, where they are more exposed, their colouration might protect them if it were not for their habit of hopping about the moment you approach them. Then, again, in the case of the small and very abundant frog, *Hyla rubella*, there is no such marked seasonal change in colour, the frogs always remaining a dull brown or grey, the bright yellow markings on the flanks and sides of the body being only visible when the animal moves. It must, however, be noted that these frogs remain close to the water's side on the sandy or, in wet weather, muddy bank, and usually shelter under stones.

The impression which is left upon one after collecting these Central Australian animals in both the dry season, when they are dull coloured and in the wet season, when they are brightly coloured, is that the often remarkable change in colouration is of very little service so far as protection is concerned, even when the change in colour is such as to produce a general resemblance between the colour of the animal and that of its surroundings; whilst in certain cases, such as that of the lizard *Amphibolurus pictus*, the brighter colours render the animal more conspicuous to human eyes and presumably to such enemies as the snakes, who certainly feed upon it.

It is, further not perhaps without interest to note that the change from a dull to a brilliant colouration takes place at or about the breeding season in the case of the frogs and lizards, but that this change, which is really just as striking as in that of many birds, has nothing whatever to do with the choice of partners.

Sometime, as in the frogs and certain lizards (such as *Amphibolurus reticulatus*) it affects equally the male and female, while in others (such as *Amphibolurus pictus* and *A. maculatus*) the male is more affected than the female.

This change in colouration actually takes place quite apart from and indeed reaches its highest development *after* pairing has taken place. What happens in the case of the Central Australian frogs and lizards is that the moment the rain falls the animals become active—the frogs come out of their hiding places—and at once pairing takes place. Every animal sets to work to feed and to reproduce its species, and in this state of general activity both male and female rapidly, but independently of, and as before said, subsequently to, pairing assumes its brightest colours.

In the warm damp ground the seeds rapidly germinate. In a wonderfully short time the bare loamy plain and even the stony gibber field becomes green with herbage; caterpillars and adult insects appear in myriads, frogs and lizards feed upon the insects whilst birds and snakes devour the frogs and lizards.

A study of the Central Australian fauna leads one to the following main conclusions with regard to colouration:—

(1) That in the dry season when food is scarce and the sum total of activities is at the lowest point, the various animals such as frogs and lizards are dull coloured, but that this dull colouration has not of necessity (as in the case of *Amphibolurus barbatus*) any definite relation to the environment, though it is often in general accord with it.

(2) That in the rainy season when food is plentiful and the sum total of the activities is at the highest point, various animals are highly coloured, but that this often brilliant colouration has nothing to do either with choice of partners (reaching its climax after pairing has taken place) or with protective colouration— sometimes even it renders the animal more conspicuous.

Many animals remain under shelter during the heat of the day; along the grassy flats kangaroos may be seen feeding, and on the Porcupine sandhills the Rat-kangaroos (*Bettongia lesueuri*) are constantly dodging in and out amongst the tussocks. The Jew lizard (*Amphibolurus barbatus*) is often seen sunning itself, and other allied species dart into their holes when disturbed. There is a great contrast in this respect between different lizards, and it is the Skinks which appear to be most susceptible to heat. One day in summer, out amongst the hot sand in the bed of the Finke, where Mr. Byrne and myself were camped, the blacks came up with a number of lizards, and amongst them a fine specimen of *Tiliqua occipitalis*. Having my hands full of specimens, I asked a blackfellow to look after it and not to let it escape, when to my surprise he simply put it down on the hot sand. It was perfectly alive when put down, having been captured in its hole, and when placed on the ground it began to travel at some rate, but after going five yards its movements became slower and before ten yards had been traversed they ceased and the animal was quite dead—simply apparently baked to death by contact with the hot sand.

About half a mile to the north of Charlotte Waters Station lies the Coglin Creek, on which by the side of a water-hole we were camped. Twelve miles away to the east is the main channel of the Finke, where, as a general rule, the waters in the rainy season spread out and are lost amongst the sandhills, though during heavy floods they may flow further south to join those of the Macumba and so perhaps swell the streams flowing into the north of Lake Eyre.

Leaving our camp on May 15th, we travelled northwards still following the telegraph line. Across the creek the country changes the stony gibber plains giving place to undulating sandy country covered with a scrub of Acacias principally Gidea with Mulga and *A. ulicina*, the latter very prickly with its aborted branchlets which have become modified into thorns.

Our black boy showed us the root of a tree (*Leschenhaultia divaricata*) which the natives put into the fire and then scrape so as to obtain a resinous material which can be used for fastening pieces of flint on to the handles of spear-throwers, etc., though in all the implements which we saw it was the resin obtained from the Porcupine grass (*Triodia*) which was used for this purpose.

There were just a few tussocks of Porcupine grass about, but we were not as yet in the true Porcupine country.

On the sand were the little crater-like pits and tracks of ant lions (Myrmeleon). The way to find the animal during the day-time is to follow up a track leading away from a crater until it comes to a sudden stop, which indicates that here the larva is at rest an inch or two beneath the sand. Only rarely during day-time can they be found at the bottom of the little craters, which are probably used at night. At first we had searched unsuccessfully below them, but a black boy on being asked to show us where the ant lion "sat down," as he called it, at once started away from the crater and followed up the track which is a very distinct groove on the surface made by the animal as it drags its body along. The aid of the blacks is simply indispensable in procuring specimens, sometimes they are at a loss, but very seldom, and as a general rule not only recognise each individual track, but from the appearance of the marks at the mouth of a burrow, will at once tell you whether the animal is in it or not.

For miles ahead from any slight rise we could see the track looking like a clean cut line in the scrub rising and falling over the low sand ridges. We camped for the night not far to the east of Mount Daniel, the highest point of a low lying range up which in the morning Messrs. Horn, Watt and Winnecke rode to endeavour if possible to gain some idea of the country away to the West in the direction of the Ayers Range, which according to the first plans drawn up it had been intended that we should visit. The rest of the party crossed a stony ridge covered with Giddea scrub and came down into the valley of the Goyder, where we camped by the side of a well dug in the middle of the channel, which was of course quite dry save for a small water-hole. As a general rule water may be obtained in these sandy beds by sinking to a depth of from thirty to forty feet, though to obtain anything like a permanent supply they must be sunk to a greater depth than this, and the South Australian Government to secure a water supply for parties traversing the overland track has sunk a series of these wells at intervals, without the existence of which it would, in seasons of drought, be very difficult to cross the country.

A permanent water supply is also being obtained from artesian bores. By the railway side at Coward Springs one of these has been sunk, and from its mouth the water rises to a height of some fifteen feet. Another has been sunk at Oodnadatta and a third is now being sunk by the side of the Hamilton, some sixty miles north of Oodnadatta; when this is complete the line will be continued northwards by one in the neighbourhood of Charlotte Waters. For stock purposes these wells will be of the greatest service. The grass which thickly clothes the loam plains of the centre during a good season is apparently peculiarly well adapted for stock feeding, and is readily eaten even when it appears to be perfectly dried up, but the great difficulty is the entire absence perhaps for months at a time of surface water, so that these wells will serve as centres from which cattle can work back over wide areas of country which would otherwise be incapable of carrying them during dry seasons.

On our way we had halted by the side of a dry clay-pan and had obtained several specimens of the water-holding frog (*Chiroleptes platycephalus*) æstivating in its burrow at the base of a Chenopodium shrub.

By the Goyder we spelled for a day, and were glad of the opportunity to do so as our specimens needed careful packing, and we had also the opportunity of a few hours collecting. The banks of the stream were bordered about three miles away from our camp by low cliffs of unconformably stratified sandstone, from the top of which we obtained an extensive view over the scrub in all directions. To the south-east lay the terraced Mount Daniel ranges and to the north-west Mount Townsend, a single, flat-topped hill with a well-marked outlier, broke the otherwise level line of Mulga scrub stretching away to the horizon.

Along the sandy bed of the river fine red-gums (*Eucalyptus rostrata*) were growing as usual, and here and there were patches of Porcupine grass. Setting fire to heaps of *débris*, we dislodged numerous lizards, amongst which the most interesting was the very agile, thin-bodied *Physignathus longirostris*. The only other representative of this genus in Australia is found in the well-watered parts of Eastern Australia, from Queensland in the north to the very south of Victoria. In these coastal districts it is semi-aquatic in its habits, spending its time on logs in and by the water side, from which, when disturbed, it rapidly dives. It is perhaps worth noticing that this genus is represented in the steppe fauna, and that in the central area it is met with along the dry creek beds, which may be taken as indicating that at one time, like its close ally, *P. lesueurii*, it also was semi-aquatic in its habits, but that in course of time it has been able to accommodate itself to a dry climate, though it still reveals its original habit in following up the creek beds.

At the Goyder was secured the first specimen of the Western black cockatoo (*Calyptorhynchus stellulatus*), which does not appear to extend further south than this. Subsequently large flocks of it were often seen, especially in the neighbourhood of water-holes amongst the ranges. The northern form (*C. macrorhynchus*) does not apparently come down into the centre of the continent.

It was very evident that the Goyder River had not run for a very long time and that therefore the country out to the west, which would have to be traversed in order to reach the Ayers Range, would be extremely dry and barren, and probably useless as a collecting ground; so, after some discussion, it was determined that we should push on northwards towards the James Range.

As the camel team travelled slowly, it was arranged that Messrs. Winnecke and Watt should make a *détour* to the west and rejoin the main party on the Lilla Creek. Accordingly we separated for a few days. The main party went on across country towards the Finke at Crown Point. After some miles' travelling through the usual scrub we came to the brow of a small escarpment forming the southern boundary of the Finke valley, the river here running almost due east and west. Away in the distance the course of the river could be distinguished by its belt of green gum trees, which extended from the point in the far north where it passed through a gap in the flat-topped hills at Crown Point.

Passing down through a picturesque defile in the Desert Sandstone escarpment, we came into the broad plains of the Finke valley, and following this up for some miles, camped close to Crown Point at the base of a line of yellow sandstone cliffs some thirty feet high.

CHAPTER III.

The Lower Steppes.

From the Finke River to the James Range.

Discovery and naming of the Finke by McDouall Stuart, in 1860—View of the Finke Valley—Cunningham Gap and Crown Point—Camp of Blacks—Their life in Camp—Corrobborees—Two important forms, ordinary and sacred—Churriia, sacred Stones and Sticks—Organisation of the Tribe—The way in which they prepare for an ordinary Corrobboree—Usual Ornaments, Weapons, and Implements—Women Mourning—Collecting amongst the Sandhills—Pyrameis kershawi and Danais petilia—Scorpions—Deaf Adder—Occurrence and Habits of Limnodynastes ornatus—Two Types of Burrowing Frogs in Central Australia—Departure from Crown Point—Reach the Lilla Creek—Meet Messrs. Watt and Winnecke at the Horse Shoe Bend on the Finke—The Horn Range—Social Caterpillar Cases on Eucalyptus microthera and Acacias—Various case Moths—Description of the Scrub—Camp at Idracowra—Determine upon Future Plans—Return of Mr. Horn to Adelaide—Visit to Chambers Pillar—Sandhills—Desert Oaks—Description of the Pillar—Myth of the Blacks to account for the Pillar—Nature and Formation of Water-holes along the Rivers—Sudden appearance of Floods in parts where no Rain has fallen—Presence of Fish in the Water-hole—No Fish in Central Australia known to have taken on the habit of Protopterus, the Mud Fish—Notoryctes typhlops, the Marsupial Mole—Is Notoryctes a form specially modified since climatic conditions became changed in the Central area, or is it the remnant of a once more widely dispersed form?—Departure from Idracowra—Cross the Palmer River and reach Henbury—Waterpool at Henbury—The Bony Bream, Chatoessus horni—Chandler Range and the Ceremonial Stone, Antiarra—Collecting amongst the Blacks Camped at Henbury—Leave Henbury—Eucalyptus gamophylla—Large Spider Webs in the Scrub—Running Waters on the Finke—Fresh Water Crayfish—Reach Illamurta in the James Range and pass out of the Desert Sandstone Area.

IT was a little to the north of the point at which we were camped that more than thirty years ago Stuart in his overland journey first struck the river course and named it the Finke. He must evidently have passed through during a fairly good season, as he says*—"I sent Kekwick to examine the Creek that I saw coming from the north. He says that there is plenty of water to serve our purpose. The creek is very large, with the finest gum trees we have yet seen, all sizes and heights. This seems to be a favourite place for the blacks to camp as there are eleven worleys in one encampment. We saw here a number of new parrots, the black cockatoo and numerous other birds. The creek runs over a space of two miles, coming from the west, the bed is sandy. The creek I have named the Finke after William Finke, Esq., of Adelaide, my sincere and tried friend and one of the liberal supporters of the different expeditions I have had the honour to lead."

The Finke River or the Larapinta as the natives call it is, despite its size, a typical Central Australian river course. In dry seasons, that is for the greater part of each year and sometimes for more than a year at a time, it may be said to be perfectly dry save for one or two deeper pools along its course across the Lower

---

* Journal of John McDouall Stuart, 1864, p. 140.

LAKE AMADEUS.

CROWN POINT.

Steppes, as for example at Henbury, and the rocky pools amongst the ranges where it and its tributaries take their rise. It flows in a generally south east direction and receives on the east two large tributaries the Ellery and the Hugh and on the west the Palmer, the Lilla and the Goyder.

It drains an area which, it is estimated, cannot be less than eighty thousand square miles, and which has, roughly speaking, the form of a triangle, the base of which is formed by the main McDonnell Range extending from 132° E. to 134½° E. A line drawn from either end of this to the northern margin of Lake Eyre will enclose the greater part of the Finke Basin. A reference to the map will show that in reality it extends somewhat further out beyond the line forming the south-western boundary of this triangle.

From the top of the cliff at the base of which we were camped the view was one very characteristic of many parts of Central Australia—that is of the Lower Steppes—over which extends the great Upper Cretaceous sandstone plain.

Just below us the river swept round in a big curve towards the east, its bed was in parts upwards of a quarter of a mile wide and simply a sheet of white sand without a trace of water. In the bed itself and forming a fringe to it the red gums (*E. rostrata*) grew with their white trunks shining brightly in the sunlight. Beyond them on the land liable to be flooded during very heavy rains grew the swamp gums or box trees (*E. microtheca*), and behind these was the undulating sandhill country covered with thin scrub with darker looking patches where the Mulga was more dense.

Three miles to the north of us the river was running from north to south through the Cunningham Gap which pierced a long range of the usual flat topped hills between two and three hundred feet high, which ran east and west. An outlier close to the western bank of the river stood out by itself, and has, from its shape, given the name of Crown Point to this spot.

This outlier is seen in the accompanying illustration (Plate 5), which is reproduced from a photograph taken with a telephotographic lens at a distance of three miles from Crown Point. Around the base of the hill is a thick fringe of river gums (*Eucalyptus rostrata*), the river bed itself lying just to the right of the part represented in the photograph. The latter shows very clearly the level capping of Desert Sandstone which overlies the softer and more friable sandstone beneath.

Everything was as dry as usual with scarcely a sign of animal life except for the crows which followed the camp everywhere, and of course ants innumerable.

On the flats surrounding the rivers were low lying shrubs and clumps of Spinifex[1] grass. In the *débris* were a few beetles and the dead shells of various land and water snails carried down by the flood from their living places amongst the ranges of the Higher Steppes.

All these rivers in the central area are liable to sudden floods. A great downpour will perhaps occur in the ranges; from the smooth rocks and the hard baked ground the water rushes in torrents into the river channels and descends at times with scarcely any warning from the highlands in the centre on to the plains to the south. For a short time the river channel is far too small to carry away the great mass of water, which spreads over all the flats along the bank. Great stretches of country previously impassable because of lack of water now become impassable from flood. Very rapidly, with the cessation of the downpour, the water is withdrawn within the limits of the river channel, where it is soon absorbed by the sand and all that is left are heaps of *débris*, new channels cut through the scrub and scattered water-holes. A rich crop of grass springs up and for perhaps a few months the wilderness is habitable, but sooner or later everything becomes as parched and dry as before.

At Crown Point, where we received much kindness from Mr. and Mrs. Ross, who live there in charge of the station, we spelled for a few days, collecting, photographing, and spending some time amongst a camp of blacks. There were a considerable number of them camped on the opposite side of the river to ourselves. The main camp was made up of a great number of smaller ones, the centre of each being, of course, a small fire. By way of habitation these blacks make at most a small wurley of branches, but as the weather was warm and fine they had not troubled to do this. Unlike those of many other parts of Australia, they never appear to make or wear any clothing, which is all the more strange as wallaby and kangaroos can often be caught, and the nights in winter are bitterly cold. The men tie a girdle, made from the hair of their mother-in-law, round the waist, and have their own well pulled back and usually cut away from the forehead, over which runs a band often made of opossum fur-string whitened with calcined gypsum. The hair behind is matted together with grease and red ochre and tied round with opossum fur string. Round the neck and arm may be worn an armlet or necklet, also made of hair and smooth with grease and red ochre, and a "figleaf," often made of the white tips of the tails of the rabbit-bandicoot (*Peragale lagotis*) or of opossum fur-string, is worn as an ornament, whilst the septum of the

---

[1] That is the true Spinifex grass (*Spinifex paradoxus*) and not the Porcupine grass (*Triodia* sp.) which is often, but erroneously, spoken of as spinifex.

nose is pierced, and through the opening made a long piece of bone—perhaps a foot in length and with a Peragale tail-tip stuck into one end—is often worn.

Except for perhaps an armlet or necklet, the women have usually no ornaments or dress.

When preparing for a corrobboree they ornament themselves—that is, the men do, as the adornment of the body is almost entirely confined to the male sex—with patterns made by stripes and spots of white, red, yellow and pink.

The term corrobboree is usually applied indiscriminately by white people to any one of the so-called dances of the aborigines ; but there are in reality at least two very distinct classes of corrobborees or, as they are called by the McDonnell blacks themselves, "quapara." One set may be called ordinary corrobborees, such as are held at any time, and which women and children may watch ; but in addition to these there is another and a very distinct series, which may be spoken of as sacred quapara, which no woman or child is permitted to see, and which are intimately connected with certain Totemic subdivisions of the tribe, members of which alone can take part in them, though members of others, provided they have undergone the ceremonies admitting them to manhood, are allowed to watch wholly or in part. Intimately associated with these are the sacred stones and sticks which have been referred to in the Anthropological section by Mr. Gillen and Dr. Stirling. The sacred nature of the implement resembling the toy commonly known as a "bull-roarer" is well known. It consists of a small flattened piece of wood, usually pointed at each end and with a hole bored through one to which a string can be fastened, the roaring or humming sound being made by the vibration of the latter when it is tightly stretched by having one end held in the hand while the bull-roarer attached to the other is rapidly whirled round.

No woman or child is ever allowed to see one of these, and should one be caught sight of by accident and the fact be known to the men, the punishment in the natural condition of the aborigines would be death, or at least blinding by means of a fire-stick. These implements, which, according to Mr. Gillen, are known as "churiña," are very highly prized and regarded as sacred. Stone ones are still more valuable and sacred than the wooden ones, which are usually spoken of as "Irula," the patterns on which are copied from the older stones, the history and origin of which are lost in the dim past.

Each division of the tribe has a certain number of Churiña, which are stored up in spots known only to the older men, or, if the locality of the store be known to the women, the latter are very careful, on penalty of severe punishment, not to go anywhere near to them. Sometimes an elder man will carry about on his person, concealed from view, one of these Churiña. It was evidently one of these stores the finding and contents of which have been described by Dr. Stirling in the Anthropological section.

The sacred ceremonies, or quapara, some of which no white man, unless, like Mr. Gillen, he has gained the most perfect confidence of the blacks, is allowed to see, and which are so jealously guarded that the ordinary white man living amongst the blacks would have no idea of their existence, are, as before said, intimately associated with these Churiña and with certain Totemic subdivisions of the four classes or phratries—Panunga, Pultharra, Purula, Kumarra—into which the Arunta Tribe is divided.

Mr. Gillen has described two of these sacred quapara, which he found to be connected with certain members of the Panunga and Pultharra phratries at Alice Springs, and the other with certain members of the Purula and Kumarra phratries. The first of these is a ceremony the object of which is intimately associated with the promotion of the growth of the "witchetty"—that is, the grub of a large longicorn beetle, which forms a favourite food of the blacks; the other is a rain- or water-producing ceremony.

The coupling of the four phratries into two pairs—Panunga and Pultharra on the one hand, and Kumarra and Purula on the other—clearly points back to an earlier time, when, as in many Australian tribes, there were only two intermarrying divisions. When four are present, as Messrs. Fison and Howitt[*] have said, we may "reasonably conclude that these four classes were formed by subdividing two primary classes, from the fact that they are composed of two pairs of non-intermarrying classes, each pair corresponding to one of the original classes and intermarrying with the other pair." Sometimes, as for example in the Mackay tribe,[†] the names for the two original divisions exist side by side with those of the four subdivisions into which they have split, but in the Arunta these two original names seem to have entirely disappeared.

In certain tribes a further division of the four into eight groups takes place, and with it a consequent greater restriction in regard to the number of women from amongst whom the man's wife must come.

The relationships, so far as marriage is concerned, of the phratries amongst the Arunta tribe is clearly shown in the articles by Dr. Stirling and Mr. Gillen,

---

[*] Kamilaroi and Kurnai, p. 47.   [†] Id., p. 38.

and it will be noted that, in contrast to such tribes as the Urrapunna and Dieyrie, who inhabit country further south, descent is counted in the male and not in the female line.

When preparing for an ordinary corrobboree a large quantity of grass down is collected and arrayed in little piles of various colours. The white is obtained by mixing it with powdered and calcined gypsum, the red with red ochre and the yellow with yellow ochre, while a pink colour is often also made by putting in less red or mixing the red and white together so that the down is just tinged. Once when wandering through the scrub at Tempe Downs I came across a party of some twelve men preparing for a corrobboree to be held in the evening. They were sitting down in a small cleared space. First of all conical helmets were made out of Cassia twigs bound together with opossum fur-string so that the point was about two feet or eighteen inches above the crown of the head on to which the broad end fitted tightly. Then they sat down in pairs, two men opposite to each other, with the requisite amount of coloured down in little heaps close at hand. Blood was drawn into the concavity of a spear thrower to serve, when congealed, as a gum with which to attach the down. As a general rule the blood is obtained by cutting a vein in the arm with a sharp flint or a piece of glass if such can be secured, but in this instance it was all obtained by probing the sub-incised urethra with a sharp, pointed stick.

Then each man took a short stick with a little opossum fur string twisted round one end so as to form a brush, dipped this into the blood and smeared it over the place to which he wished to attach the grass down on to the helmet, face or body of his friend sitting opposite to him. In some cases (as shown in the illustrations of the Anthropological section) the pattern thus formed is a very regular symmetrical one, in others it is asymmetrical. Very often the whole front of the helmet and the face, as far down as the mouth, is covered with a regular solid pattern of down which just leaves two circular patches in the centre of each of which is an eye.

The pattern may be continued right on down the body and along the legs and arms and very frequently (depending of course upon the special corrobboree being enacted) the toilet will be completed by a tuft of eagle feathers waving from the apex of the helmet and, as in this particular instance, by anklets and armlets of little leafy twigs of the gum tree.

Whilst this preparation is going on, and it may last for hours, a low humming of a corrobboree tune is kept up, though, every now and again they burst forth into a louder refrain and then gradually sink back into a subdued and monotonous repetition of the notes as if the music were dying away in the distance.

In the evening we saw the corrobboree performed for which these blacks were thus preparing, and one of the movements in which is represented in the photograph reproduced in the Anthropological section. Sometimes the complete performance of such a corrobboree will extend over several evenings, reminding one in this respect of the long drawn-out performances of the Chinese and Japanese.

In the camp at Crown Point they were gathered together in little groups, men in one, women and children in another, the fire always the central point. Some of the men were weaving opossum hair string, others were making implements of various kinds, and others grinding Munyeru. The latter is the little black seed of *Claytonia Balonnensis*, and it is prepared by putting it on a large flat stone and then grinding it with another small flat stone held in the hands. Water is every now and then poured on, and the muddy looking mixture tumbles over the edge of the under stone into a receptacle and is then ready for eating, either raw or after roasting. Another favourite food is the bulb of an amaryllid plant called by the natives Irri-ākūra.* This the women go out and gather in hundreds. The ground Munyeru tastes as it looks, like black mud; but the Irri-ākūra is not at all bad and has a decidedly nutty flavour.

As to weapons and implements, these are comparatively few in number and usually devoid of the elaborate finish and ornamentation characteristic of those belonging to the more northern blacks. A complete account of them is given in the Anthropological section, so that only those will be mentioned here such as every blackfellow carries about with him.

The spears are of two kinds, barbed and unbarbed, the former usually made of at least two pieces carefully spliced together. The main shaft is often composed of Tecoma wood straightened by careful heating in the fire and by subsequent pressure. The point is made of hard Mulga, and to this a little recurved wooden barb is affixed by means of emu or kangaroo sinew. The unbarbed form may be of considerably greater length than the barbed one, and the rarer ones are made out of the desert oak (*Casuarina Decaisneana*). Every man carries two or three spears and a spear thrower or amēra † This is a broad hollowed-out piece of wood, tapering

---

* This proves to be *Cyperus rotundus*, having been determined from plants grown by Dr Stirling from bulbs brought down by himself to Adelaide. In other parts it is known as "Nutgrass."

† The word "wommera," so common in various other parts of Australia, is not applied to the spear-thrower in the central districts.

gradually towards the end which is held, and abruptly towards the opposite end where a little wooden point is attached by tendon. The other extremity has a round knob of resinous material obtained from the porcupine grass, and into it a roughly sharpened flint is often attached and used for cutting purposes as for example to trim down the rough surfaces of a spear, or to cut open the body of a kangaroo.

Boomerangs of various sizes are made; the larger ones are very heavy and simply used for fighting at close quarters, the smaller, flatter ones are thrown, but they do not appear to have any so made that they can return to the thrower. Shields are made out of light wood such as that of the Bean tree (Erythrina); they are perhaps two feet six inches in length, very thick, with a strongly convex outer side and a slightly concave inner one, in the middle of which a cavity is made leaving a bar, running across in the direction of the length of the shield, which can be grasped by the hand. As the Bean tree does not grow so far south as this, these shields have to be traded from one part to another.

The women are usually provided with pitchis, which are receptacles hollowed out of wood and used for carrying food such as grass seeds or Irri-akura bulbs. They vary considerably in size and form, and some are made out of light wood like that of the Bean tree, and others out of heavier wood such as Mulga.

Amongst the women and lubras were one or two in deep mourning, which was indicated by the fact that the hair ringlets were stiff with white gypsum, whilst a band of the same was plastered over the bridge of the nose and on the cheeks and forehead.

These natives belong to the Arunta tribe, which occupies a large tract of land stretching from the Macumba Creek in the south to about seventy miles north of Alice Springs. Westwards it extends to Hermannsburg, and its eastward extension is not completely known. At Alice Springs it spreads out for about a hundred miles to the east of the telegraph line. Very often the men used to describe themselves as Larapinta blacks, from the native name of the Finke River, which drains a considerable part of the country which they occupy.

Many of the men were well built, though, as usual, the legs were the weak point. The tallest one measured by Dr. Stirling was 5 feet $9\frac{5}{8}$ inches in height, and the average of ten of them was just under 5 feet 6 inches. The women are decidedly shorter, the average of ten of them being only 5 feet $0\frac{3}{4}$ inches. The men, with their long, flowing beards and hair cut off their foreheads and the rest tied back with a white band, often looked very patriarchal, an appearance frequently enhanced by their dignified bearing, though at times the presence of a bone perhaps a foot in length stuck through a hole in the nasal septum and ornamented at one end with a Peragale tail, detracted, to a certain extent, from the dignified appearance of the wearer. So long as food is plentiful they are perfectly happy and contented, their disposition being just like that of light-hearted children who have no idea of anything beyond the enjoyment of the present moment.

As usual, the harder work is done by the women, who have, in addition to looking after the children, to go out in search of animals such as lizards and of the grass seeds and bulbs, which form staple articles of food, the men procuring the larger animals, such as wallabies and occasionally kangaroos and emus. To their children they are very indulgent, the young boys being especially well treated, though in occasional fits of anger acts of cruelty may be performed. Anything given to them is at once shared with other members of the camp. If you give a black, say, a woollen shirt you will find him wearing it one day, his wife will be adorned with it the next time you meet her and perhaps some friend will be wearing it the day after. At the same time, they have a distinct idea of private property. In camp, for example, each man will have his own belongings and such as he is not carrying about with him will be left close to his fire quite unprotected, in the certain knowledge that, so far as his fellow blacks are concerned, they will not be interfered with. At the same time, it is quite recognised that if you possess, say, a spear, and a friend asks you for the loan of it, you are in duty bound to lend it. Everything has its special owner, though he may be very many miles away. Whilst a man will part with his own property he will not do so with that of anyone else when this has been lent to him. I once even had considerable difficulty in persuading a man to part with a tuft of Peragale tail-tips which belonged to his wife and on more than one occasion I could not secure things because they had been lent to the possessor.

The question of the possession of land is a more difficult one. There is, of course, no doubt that they have no idea whatever of any tract of country as belonging to any individual; but, on the other hand, they have a very distinct idea that certain tracts of land, and the right to inhabit and hunt over them, belong to particular groups. Within the limits of the Arunta tribe, for example, there are subdivisions occupying well-defined districts. A man belonging to the Arunta at Alice Springs coming down south to Charlotte Waters, for example, is regarded as a guest and as such is allowed certain privileges. Thus not only have the tribes such as the Arunta, lands which are regarded as belonging to them, but there are divisions of the tribes which in the same way are regarded definitely as owning special tracts of country, the boundaries of which are well defined.

Whilst at Crown Point a considerable amount of collecting was done. Amongst the sandhills behind the camp were numerous lizards such as the strange *Moloch horridus*, the bright yellow, orange, red and black of which render it in life very different in appearance from the bleached specimens of museum cases. The Jew lizard (*Amphibolurus barbatus*) was often seen, some of them being of a curious brick red colour similar to that of many of the sandhills amongst which they lived. There are perhaps no animals, amongst land forms, of which museum specimens give so poor an idea, so far as colour and shape of the body are concerned, as frogs and lizards. Both in brilliant, and often also in delicate colouration, many of the Central Australian ones cannot be excelled. A reference to the plates illustrating the article on lizards in the zoological section of the work, where the natural colours have been most carefully represented by Mr. Wendel, will serve to show how inadequate an idea the ordinary museum specimen conveys of the real appearance of the living animal.

It was not the right time of the year to secure many butterflies and moths, but two species were very common everywhere. One was the Australian "painted lady" (*Pyrameis cardui*, var. *kershawii*), which has been described by Sir Frederick McCoy as specifically distinct from its close ally, the European form (*Pyrameis cardui*). Other writers, however, such as Kirby,* regard the European and Australasian forms as "hardly to be considered distinct." The chief distinction, apart from size, between the two lies in the fact that the three black spots on the hind wing have blue centres in the Australasian species.

The other is an introduced form *Danais petilia* a pretty black and yellow insect, feeding on a Labiate plant (*Cynanchum floribundum*). In other parts of Australia such as Queensland, another species of the same genus (*Danais erripus*), also introduced, is met with feeding upon an introduced Labiate plant.

The burrows of a scorpion were very common, and its tracks leading into a hole in the sand just big enough for it to crawl into were very well marked ones and easily recognisable when once the blacks had told you what animal they were made by, for it was never seen during daytime in the open. The burrow goes down to a depth of three or four feet. We very rarely found the animal under stones or logs as, unlike those met with in the coastal districts such as Gippsland, they seem to generally make burrows in the sandy ground. In the bed of the Finke during summer time I found them crawling about at night on our camping ground.

---

* Handbook of the order Lepidoptera, Part I., Butterflies, vol. i., Allen's Naturalists' Library, 1891, p. 92.

In the sand by the river bank, a foot or two below the surface, was a beautiful little black and orange banded snake (*Vermicella annulata*) some six or eight inches long of which, for some reason, the blacks were very frightened. It does not however do to trust implicity to the natural history instincts of the natives. One day at Charlotte Waters, during my second visit, they brought in a specimen of what was evidently either the "deaf-adder" (*Acanthophis antarctica*) or another species of the same genus. It was longer and thinner in the body and more brightly coloured than the usual specimens, but had the same general appearance, and the little spine at the tail end which is distinctive of the "deaf-adder"—the most venomous of our Australian snakes. Despite this, they were positive that it was a non-venomous snake and handled it in a way in which they would not do even a Hoplocephalus. I questioned two or three of them about it but they would not alter their opinion, and yet it turned out on examination to be a true *Acanthophis antarctica*.

Right in the sand of the river-bed were every now and then the tracks of a frog. As the sand on the surface was very hot and dry I was a good deal surprised to see these, but, of course, the blacks knew all about them, and after following up the tracks of one for a few yards they came to an end at a spot where there was a little depression as if an animal had burrowed down and the soft sand had been pressed up on either side and had then slightly tumbled in towards the centre as the animal went down. A foot or so beneath the surface it was cool and slightly moist and here we came upon the frog (*Limnodynastes ornatus*). Its body is about two inches in length, the ground colour being a silvery grey with splotches varying in colour from dark grey to umber. There is always present a light lyre shaped patch on the hinder part of the head, the two arms of the lyre stretching forward one on to each eyelid. On the upper surface of the body and limbs are little dots of salmon pink colour surrounded by small dark circles. Sometimes the dark markings are so indistinct that the whole body has a silver grey appearance but the pattern, however feebly indicated, is always one which can be derived from a well-marked, dark specimen.

The hind foot is strongly webbed and has the shovel-shaped tubercle which is so characteristic of burrowing frogs. Sometimes the body is to a certain extent swollen out with water which can be pressed out through the cloaca; but this is nothing like so striking a feature as in the case of the clay-pan or water-holding frog previously described (*Chiroleptes platycephalus*).

The stomach contained beetles which had evidently been caught on the sand, the animal emerging from its hiding place during the night when everything is cool. It then hops about in search of food, and at the approach of day burrows down into the cool damp sand below the surface.

There are thus two types of burrowing frogs in Central Australia—one the clay-pan frog, forming a permanent burrow; the other, the river-bed frog, forming temporary burrows.

The same species as the latter is found in Queensland and New South Wales, but so far as is yet known, it has only adopted this burrowing habit in Central Australia and with that may be associated its strongly webbed feet which are very unlike the typical examples of the genus to which it belongs.

On May 19th, we started from Crown Point and left the telegraph line to the East of us. It was more than two months before we struck it again near to Alice Springs.

Travelling West through the usual scrubby country we reached the Lilla Creek which flows into the Finke from the desolate barren country out to the West. After reaching camp close to the Lilla Creek we were surprised to see our black boy Harry who had gone out with Messrs. Winnecke and Watt. He had ridden across country to try and intercept us with a note saying that the two latter had changed their plans and would meet us on the Finke, near to a place where, hemmed in by a semicircular escarpment of high sandstone cliffs, it sweeps round what is known, from its shape, as the Horse Shoe Bend, and to the blacks as Engoordina. This necessitated a slight change in our plans as we had arranged to meet higher up the Lilla, and so crossing the latter instead of following it up westwards, we struck the Finke close to Mount Musgrave, a curious pyramidal peak rising from bare stony plains.

We found that Messrs. Winnecke and Watt had first followed up the Goyder for some distance finding no water. Then they had struck across north-west into some hills to which the name of Horn Range was given, and then crossing the Lilla Creek which was, like the Goyder, perfectly dry, they had travelled north-wards to the Finke. On a hill lying some forty-five miles north-west of the junction of the Lilla and Finke they had found a Silurian formation containing fossils—a find of some importance as this lies a considerable distance to the south of the previously known Silurian formation of the James Range, from which it is separated by a wide tract of Desert Sandstone country.

From Engoordina we travelled north-west across country so as to reach Idra-cowra, the course of the Finke here forming two sides of a triangle of which our track formed the third and south-western side.

The country was sometimes slightly undulating with reddish sandy soil covered with scrub, at others it rose to bare stony plains on which grew perhaps a few stunted Mulgas and low blue-white salt-bushes.

All along the creeks were the usual red gums in the sand bed, and on the banks and then beyond these was a fringe, varying in width, of swamp gums or

box trees (*Eucalyptus microtheca*). Everywhere along our route from Oodnadatta in the south to Alice Springs in the north, these swamp gums especially were infested by a particular kind of caterpillar. They live socially in big, bag-like cases attached to the branches, one of which is represented in the accompanying illustration. A single case will sometimes measure as much as two feet in length, and will contain perhaps a hundred or more caterpillars whose excrement mixed with hair from their bodies fills the case and is of the most irritating nature if it falls on the skin. They are most frequent on the swamp gums—they hardly touch the red gums at all—but are also found on Acacias and Cassias. On one Acacia (*A. salicina*) I counted no fewer than fifty-seven of various sizes. From the case a track of web-like material can sometimes be seen running down the trunk to the ground. This calls to mind the ladder-like track made by the caterpillar of such a case-moth as *Metura elongata*. When the caterpillar of the latter climbs a wall or tree trunk it lays down a series of lines of web which are arranged one above the other just like the rungs of a ladder, and to which while climbing it clings with the claw-like tips of its feet. The track must be made by the caterpillars who come out of the case to feed and are said to walk about in long processions. Finally, after stripping the trees of their leaves (in August when we returned scarcely a leaf was to be seen on the trees on which the empty, broken cases were swinging about before the wind) they all come out of the case and burrow into the ground and there chrysalate. Unfortunately we had no means of determining the insect to which the caterpillar belongs.* Travelling over the same ground six months later, the trees were all once more green, and there was not a trace of the cases to be seen.

Many caterpillars live in colonies, and some spin a web inclosing leaves and twigs and so forming a nest or case for themselves, but in this instance the leaves are not utilised, and the big bag hangs in the most prominent position, attached usually to the smaller twigs at the ends of the gum tree branches. Nothing will interfere with them owing to the exceedingly irritating nature of the excreta which they contain, and the blacks believe that if this falls on your face you are sure to be blinded. Serious results are known to have followed the tumbling down of a case on to a white man sleeping under the gum tree from which it fell.

The web is strong but nothing like so tough as that of the two or three kinds of "case-moths" found in the scrub. In each of these, which is a small often tubular bag one or two inches in length, there lives only a single caterpillar. The latter carries its case about with it while it feeds, and finally turns into the chrysalis inside it, the male moth comes out at the lower end but the female is said never to leave it,† never in fact developing any but the most rudimentary wings and appendages. There are thus several distinct kinds of protective cases or houses made by caterpillars, some of which are concerned with single caterpillars, and others with social ones. The simplest is the irregular webbing which serves to fold over the edge of a leaf, and which is made by, and incloses only, a single larva. There are gradations in structure between this, and the most highly developed form of house made by a single animal such as the common case-moth (*Metura elongata*).

Of cases inhabited by more than one larva the simpler are again made by enclosing with web, more or less regularly, leaves and twigs, whilst the most highly developed is perhaps this large bag-case found in Central Australia, which may serve as the hiding place for more than a hundred caterpillars. These cases or houses must not be confused with the true cocoons that is the structures made by

---

* It is probably a species of the genus Teara. † There is some doubt about this.

the caterpillar just when it begins to chrysalate, and in which the chrysalis lies. As we have seen the social Central Australian caterpillars come out of their common case, each one goes into the ground and there chrysalates. In the case of Metura the house is used first by the caterpillar, and subsequently by the chrysalis, but even here it is not to be regarded as equivalent to, or even taking the place of, the cocoon, as the larva when passing into the chrysalis stage makes a rough kind of cocoon for itself inside the outer case. The latter, in fact, may always be regarded as a structure connected with the larva or caterpillar stage, and the cocoon as simply connected with the pupa or chrysalis stage.

The scrub over all this part of the country was characteristic of the general scrub of the steppes in the Finke basin south of the ranges—that is one which is formed of a mixture of various kinds of plants, and not made up mainly of one kind as in the case of Gidden, Mulga or Mallee scrub.

Amongst the shrubs and trees four genera were dominant—Eucalypts, Acacias, Cassias, and Eremophilas.

The Eucalypts away from the river beds are usually of the species known as mallee gums—with no tall central stem, but with a number of thin branches springing from a stem which projects, at most, for only a short distance above the ground. They grow in patches, but are nothing like so plentiful in this as in other parts of Australia, where dense mallee scrub will stretch monotonously over mile after mile of country.

The Acacias vary in size from a few feet to twenty or thirty in height; as a general rule they have thin harsh petioles serving as leaves; their general tint as in the case of the Mulga is a dull depressing olive green, though at times, *Acacia salicina* will grow into a tree with denser and greener foliage, sometimes hanging gracefully down as if it were a weeping willow. This is really the most attractive tree in the scrub, and often the only one beneath which any shade can be obtained when the sun is shining in a brilliant cloudless sky, and the strong light is reflected from the hard yellow ground.

The Cassias only grow into good sized shrubs, perhaps at most, six or eight feet high. Their leaves as a general rule are thin and their branches straight and wiry, but still they often form green patches which, covered with masses of yellow blossom, serve to brighten the dull scrub.

Often for miles together the greater number of the Cassias are dead, and their thin wiry branches all springing together from a short stem close above the ground curve over until they meet each other at the top, and look as if they had been purposely tied up into a bundle swelling out in the middle.

The Eremophilas as a general rule form smaller shrubs than the Cassias, their leaves are nothing like so wiry, and their branches are not so straight, so that they are thicker and more bushy in appearance.

All the Cassia flowers seem to be yellow, but the tubular corollas of the Eremophilas are of various tints, blue, pink, purple, white or yellowish, with dark red spots.

In addition there are, every now and then, patches of rugged stemmed Hakeas with stiff spike-like leaves, or of Grevilleas—the "silky oaks" with their grey-white foliage. Now and again is a Codonocarpus, the "native poplar" with light green leathery leaves, or a big, bright green bush of Dodonea with shining viscid leaves, or a stiff broom-like Templetonia.

Many of the trees—Acacias, Eucalypts, Grevilleas—are studded with clumps of Loranthus—the Australian Mistletoe, with green, red or yellowish flowers and white or red berries. Occasionally a single Acacia will have more than one species growing on it, but on the other hand some species will be found confined principally to one tree as for example *Loranthus gibberulus* to the Grevilleas.

The ground is hard, yellow and sandy with tufts of withered grass. Now and again it is brightened with a patch of "everlastings"—yellow Helichrysums or beautiful white, purple and pink Ptilotus in full flower—but as a general rule the sharp shadows of the thin scrub fall only on bare ground studded with endless little ant hills.

Such is the general character of the scrub through which we travelled day after day, at times passing along flats by the creek beds where the vegetation was a little less parched up than usual, or rising on to upland plains with salt bush, and great spherical masses of Salsola, which, when withered up, are easily torn from their roots and are carried away, bounding like yellow balls before the strong south east winds which prevail during the winter months. This salsolaceous vegetation is very characteristic of the stony and loamy plains of the Lower Steppes, lying between Oodnadatta on the south, and the Finke at Idracowra on the north; in this district there is only a very little Porcupine grass (Triodia sp.) to be seen; but, further north again, in the Higher Steppes, this becomes a dominant feature of the dry sandy flats and often of the hill sides extending right to the top of the highest peaks such as Mount Sonder.

As we came near to the Finke at Idracowra, and stood on the edge of the highland, bounding its broad valley on the south, we could see away to the north the track of the river indicated by its thick fringe of gum trees, and, beyond this again, over the rolling sandhills, Chambers Pillar stood out clearly against the sky.

Passing down into the broad open valley we camped beside a water-pool containing plenty of fish, and here we spelled for a day to allow some of us to go on and photograph Chambers Pillar which forms one of the most prominent land marks in this part of Central Australia.

We had now travelled over a distance of nearly two hundred and fifty miles from the head of the railway at Oodnadatta, and, at this point, Mr. Horn was unfortunately obliged to leave the party and retrace his steps to Crown Point to meet there the overland mail which, once in every six weeks, runs between Alice Springs and Oodnadatta.

The evening was spent in discussing and settling upon plans for our future movements, and a rough outline of these was drawn up. If possible it was decided that we should make for the following points in succession, Henbury, Running Waters on the Finke, Illamurta in the James Range, Petermann Creek, Tempe Downs, the Levi and George Gill Range; from the latter a party was, if possible, to go down south across Lake Amadeus to Ayers Rock, whilst the main camel team went north to Haast's Bluff at the western end of the McDonnells. The two divisions were then to meet at Mount Sonder, from which a detour was to be made south to Hermannsburg on the Finke, close to the Glen of Palms. Then passing northwards again the main McDonnell Range was to be struck about the neighbourhood of Paisley Bluff, and thence travelling eastward the whole party would reach Alice Springs. The actual route was to be determined upon by the four members of the science staff, and upon Mr. Winnecke developed the responsibility of carrying out the wishes of the scientific staff so far as he judged them to be consistent with the safety of the Expedition.

Leaving the loading camels to enjoy a well-earned day's rest amongst feed and water, four of us took our riding camels on and went northwards for ten miles over the sandhills to Chambers Pillar. These sandhills vary in height from twenty to forty feet, and run in long rolling lines from north east to south west. Each has a sloping southerly and a steep northerly side indicating the long prevalence of the strong south-easterly winds. On the south side the thin scrub covers them, but on the north each has a long bare line of bright red or yellow sand just where the slope is steepest, and no vegetation can grow. From the top of each one

as we mounted them in succession—for, unfortunately, our course lay nearly at right angles to their length—we could see the Pillar standing out against the sky above the hills which looked like great waves of sand piled up one behind the other.

We had now amongst these sandhills come into the region of the "Desert Oak" (*Casuarina Decaisneana*). Some of them reach a height of forty or fifty feet, and, growing either singly or in clumps, form a striking feature amongst the thin sparse scrub (see Plate 1). Evidently a shower of rain must have fallen recently, as there were more flowers about than usual. The shrubs of Eremophilas, Acacias and Cassias were bright with purple and yellow flowers, and on the ground were exceptionally fine white, pink and purple blossoms of Ptilotus, and patches of white and yellow Helichrysum picked out with the little blue flowers of Brunonia.

The "Desert Oaks" with their pendant, wiry twigs, which take the place of leaves, have a strangely weird appearance. The older trees, as shown in the illustration, have a main trunk with a rough bark rising to a height of perhaps twenty or even thirty feet at which point a large number of strongly developed branches are given off. The younger ones resemble nothing so much as large funeral plumes. Their outlines seen under a blazing sun are indistinct, and they give to the whole scene a curious effect of being "out of focus."

Some idea of the size and extent of the sandhills may be gained from the fact that in nine miles traverse we were continuously passing up the gradual southern slope and down the steep northern face, then across a short level flat on to the slope of another, and that in this distance we crossed some thirty-five.

At length we came to a small level stretch of land from which the pillar rises. It has the form of a tall column placed on a broad pedestal. The latter has a circumference of about five hundred yards, and a height of one hundred feet, the column itself is nearly seventy feet high, and is roughly oblong in section, one side measuring about twenty-five, and the other about fifteen yards in length.

The whole is composed of a friable sandstone capped with a layer of the same chalcedonized sandstone which forms the thin uppermost layer of all the flat-topped, Desert Sandstone hills, and has been the means of protecting the softer rock beneath. At one time the whole of the country must have been at the level of the top of the pillar; now all save this solitary column and a few remarkable turret-like peaks, forming what is called Castle Hill, a short distance away to the north, has been worn away, and the pillar stands solitary amongst the sandhills (Plate 6).

In colour it is a pale cream-yellow, except just the upper part where the oxidation of iron contained in the rock has tinted it bright red, and standing out against the blue sky above the yellow sandhills and dull green scrub it forms a striking feature in the otherwise monotonous landscape.

The blacks have a rather curious myth* to account for the origin of the pillar. They say that in what they call the Alchëringa (or as Mr. Gillen appropriately renders it the "dream times"), a certain noted warrior journeyed to the east and killing with his big stone knife all the men, he seized the women and brought them back with him to his own country. Camping for the night on this spot he and the women were transformed into stone, and it is his body which now forms the pillar, whilst the women were fashioned into the fantastic peaks grouped together to form what is now known as Castle Hill, a mile away to the north (Plate 1).

After photographing we returned late in the evening to our camp by the Finke, ready to start away in the morning to follow up the river as it came down from the James Range in the north-west.

Close to our camp there was a fair sized water-hole in the sandy bed of the river in which we secured a few fish. The water apparently remained here owing to the deposition of a thin layer of clayey material on the sand which prevented it from sinking in as it had done elsewhere, and the fish were simply existing until, in a very short time, they must perish when the water dried up. Along the course of the Finke as it meanders over the country between the James Range and its termination somewhere amongst the sandhills to the north of Lake Eyre—for only in exceptionally wet seasons can its waters reach the lake—water-holes are met with at intervals, but very few indeed, if any of them, can be regarded as permanent, and they only last for a varying length of time after the rain season.

When there is a heavy rainfall then the floods come down the channels and, in favourable spots, the water will lie on the surface while elsewhere it sinks down into the sandy bed. The gathering ground lies far away amongst the ranges up country so that not infrequently a flood will occur at Crown Point or Idracowra without there having been any rain in these parts, and when the water does come down the river channels it does so with great force and suddenness. But little warning is given of its approach, though, at times, the blacks send on the news of an approaching flood so that especially during the rainy months up country, that is in summer time, it is not really safe to camp in the sandy bed of a river however

* I am indebted to Mr. F. J. Gillen for this information

dry it is and may have been perhaps for months before. Before the morning comes the formerly dry channel may contain a roaring torrent bearing uprooted trees and shrubs along with it and spreading out far and wide over any low lying flats, and your baggage and impedimenta may be miles away down stream.

The falling of the waters is almost as rapid as the rising; as soon as the rain ceases in the gathering ground the torrents from the rocks and hard soil dwindle and dwindle until they disappear. The overflow level of the rock pools amongst the ranges is soon reached for there are no springs yielding a constant supply from waters which have been stored up amongst the rocks. At most any small supply of water oozing out of them serves only to balance the loss of evaporation so that there is no surface flow. Very rapidly, when once the supply from the high lands ceases, the water drains off, a large part fortunately sinking down into the sand where at some depth below the surface it slowly finds its way along. Where the scour has been the strongest and has washed out the sand, there for a longer or shorter period water will stand, prevented from sinking into the sand by a thin coating of the finer mud which, when the water is of some little depth and stands quietly for a short time, will fall down and give rise to a thin rather clayey film on the surface which will help to keep the water from sinking. Fortunately the conditions are favourable to the formation of this impermeable film. When the waters are at their highest the holes are scoured out in the sand; as they fall and begin to flow along sluggishly first of all the heavier sand particles are dropped, lastly there remain only the finer mud particles with which the water, now forming only a thin surface layer, is heavily charged. As the surface layer flows over and into the deeper pools, it brings into them a constant supply of fine mud until gradually the flow ceases and the mud settles at the bottom and by its means a supply of water is retained.

Amongst other things, the flood brings with it fishes from the more permanent sheltered pools amongst the ranges, and they survive as long as the water lasts. Nowhere in the central region is there any evidence of the fishes having adapted the habit of the African mud-fish (Protopterus) to enable them to withstand a dry period. There is indeed no Australian fish which can, so far as we yet know, exist if the waters be dried up, and in the central region the reason for this is not far to seek. The river beds are sandy and after digging down for some thirty feet no material is met with out of which anything like a mud case could be made.

Where the burrowing frogs are found by the side of water-holes and clay-pans on the stony table-lands further south, no fish are met with. They could only reach these water-holes in the form of eggs carried by birds, and long before they could have developed, in most cases, the water is dried up. In addition to this they would, if they were to survive, have to acquire the habit of burrowing down to a considerable depth, else (unless like the frog they absorbed a sufficient supply of water within their body) they would simply be desiccated.

The only water-holes not in the course of river channels in which, so far as at present known, fish exist are the permanent ones associated with mound springs and artesian wells. In the warm waters of one of the Dalhousie mound springs, which unfortunately we did not have the chance of visiting, fish are found though the species is not yet determined, and in the pools around the opening of the artesian bore at both Coward and Strangway's Springs, the water in the case of the former being only a few inches deep, is found a small Gobius (*G. eremius*), which is at present known to exist only in these two artificial water-pools. In the case of the bore at Coward Springs the water issues at a temperature of 95° F. There can be little doubt but that these two water-pools have been stocked with fish by means of eggs brought to them attached to the feet or feathers of some of the numerous birds which, immediately after the rainy season, appear in numbers and fly about from one water-hole to another; but where they were brought from is unknown as the species is not yet recorded as occurring in any natural water-pool.

Our camping place, Idracowra, was interesting because it was from this locality that a few years before Mr. Bishop had obtained for Dr. Stirling the larger number of specimens of that interesting marsupial *Notoryctes typhlops*. It was too dry during the time which we spent in the district to obtain specimens of the animal except one solitary one which, through the kindness of Mr. Ross of Crown Point, was secured by a black boy and brought alive into camp but it soon died, and there is apparently very little hope of their ever being brought down alive from the central region.

Since the return of the Expedition, Mr. Byrne has secured in the neighbourhood of Charlotte Waters, a considerable number, some of the more interesting points in connection with which are dealt with in the section of the work dealing with the zoological results. They live in the sandhill districts, and though not easy to capture owing to their power of rapid burrowing, still they are not perhaps quite so rare as it was at one time thought they were. During the past season which has evidently been a favourable one, between forty and fifty have been captured within a radius of thirty miles of Charlotte Waters. The blacks say that they can catch them best when there has been a fall of rain, as then their tracks are more distinct. They do not make a permanent run like a mole does, for the obvious reason that the loose sand near the surface simply falls in and obliterates the burrow as the animal passes along, so that it is necessary to be able to follow their tracks on the surface, and this the blacks always tell you that they can only do after rain. It must be confessed that this is not an altogether satisfactory explanation, but it is one always given by the blacks. The latter will follow any track up on dry or wet sand, and there can be no difficulty whatever in their detecting the track of a Notoryctes however dry the sand is. I fancy that the real explanation lies in the fact that the blacks catch the beast on the surface when they happen, by chance, to come across it, and that, for some reason, it is most frequently seen on the surface shortly after rain.

It is a curious feature about Notoryctes that though absolutely blind still it normally spends a part of its time on the surface, and the complete loss of eyes externally is, no doubt, to be associated with the fact that it is constantly burrowing in loose and often hot sand, the grains of which would, if it had eyes, be a fruitful source of irritation.

The affinities of Notoryctes to other forms of marsupials are somewhat obscure, but they evidently lie with the Dasyuridae rather than with any other family.

It is somewhat difficult to understand the remarkable modification evidently undergone by Notoryctes, whereby it has become adapted to its present mode of life. The modification in regard, for example, to its complete loss of eyes on the surface, and its shovel-shaped feet are evidently correlated, not only with its burrowing habits, but with the fact that it burrows in soft, loose, sandy country. If this modification to adapt it to a burrowing habit has taken place during the period (since Pliocene times) in which the central area of the continent has assumed its present desiccated condition, then it is somewhat difficult to understand how this one form has become so much modified in the time which has only served to produce slight modifications amongst members of other families, such as the Dasyuridae. On the other hand, and this is perhaps the most probable state of the case, Notoryctes may be the one (as far as yet known) surviving representative of a once more widely dispersed burrowing and mole-like marsupial, which, for some reason, has been left stranded in the central region and has elsewhere disappeared. It is not difficult to understand how Notoryctes—at the best an animal of rare occurrence and inhabiting only districts which are comparatively inaccessible and not yet by any means thoroughly explored, zoologically— remained for so many years unknown, until by a happy chance it was sent down to the South Australian Museum, and its existence made known by Dr. Stirling. There, perhaps, yet remains to be discovered in the remoter parts of the continent some allied burrowing marsupial, or possibly search amongst Tertiary rocks, may lead to the discovery of allied extinct forms.

After leaving Idracowra we travelled westwards, crossing and recrossing the bed of the Finke all day long as it meandered about, until at night we camped not far from the Johnston Range, which forms a bold escarpment of red sandstone with the usual level capping of Desert Sandstone.

Here we picked up a young black boy, who went on with us clothed with a thin hair girdle round his waist and a head-band. Of all the black boys with whom we met, this youth was perhaps the most loquacious and anxious to impart information. Having been recently admitted to the privileges of manhood, there was little he did not profess to know with regard to the habits and customs of his tribe, but as such knowledge is only to be gained from the elders, his information, all the more freely volunteered because it was the result of, for a blackfellow, a somewhat vivid imagination, was accepted with considerable reservations.

Crossing the junction of the Palmer and Finke, we reached the outlying station of Henbury, and were most kindly received by Messrs. Parks. The station lies by the side of a deep water-hole in the Finke, where a bar of rock crosses the stream and so has caused the formation of a deep pool which is full of water and in which, by means of a net, we caught hundreds of specimens of a fish which is known locally as the "bony bream." It is the commonest fish along the river, and proves to be a new species which has been described by Mr. Zietz under the name of *Chatoessus horni*. It has much the general appearance of a bream, with bright silvery scales and somewhat flattened body. The largest specimen secured weighed upwards of a pound, but after cooking them we found that the name "bony" was most suitable, and though it is the most abundant of all the Finke fishes, it is not really of much use as food when anything else is obtainable.

At the time of our visit there was a splendid supply of water in the deep pool, but yet even such an apparently permanent water-hole as the Henbury one is liable to be destroyed by a moderate rainfall, which will fill the river and so, bring down sand enough to fill the pool. So long as a real flood of water comes down the bed of the river, the band of rock causes a sufficient scouring out to ensure the formation of a deep pool behind the rock, but if only a moderate amount of water comes down the river, then the rock may simply act as a barrier behind which the sand grains may accumulate and fill up what is now a deep pool, in which case the water will disappear from the surface and find its way round, underground, at some spot where the rock lies lower beneath the surface.

Fortunately the rain appears to fall in sufficient quantity to always scour out this pool in front of the rocky barrier.

Henbury, where we spelled for a day, lies not far from the northern limit of the Great Cretaceous Plain forming the Lower Steppes over which we had been travelling. At Lake Eyre, close to our starting point, the land was actually below the sea level, but here we were one thousand feet above it and were close to the ancient Silurian or Ordovician mountain ranges around the base of which, much higher and more imposing then than they are now, must have washed the waves of the old Cretaceous inland sea.

The flat-topped Desert Sandstone hills which we had passed by during our journey had indicated the former level of the land, and though all was now dry and sterile, the discovery of vast remains of Diprotodons and other extinct forms at Lake Callabonna, as well as the general physiographic features of the region, have shown that between the far-off time when the land first rose above the level of the Cretaceous sea and the present time, there was an interval during which, in contrast to its present state, the land was covered partly with great fresh-water lakes and partly with rich forest growth, capable of supporting an extensive fauna such as could not possibly exist at the present time.

It was in 1836 that Mitchell, in his expedition to the Rivers Darling and Murray, first discovered the Wellington Caves and found in them the remains of a gigantic fossil marsupial, to which in 1838 Owen gave the name of *Diprotodon*. Since that time its remains have been found in many parts of the interior of Queensland, New South Wales and South Australia, and in the western parts of Victoria, occurring in formations now usually described as of Pliocene age. At the same period lived numerous other and now extinct forms which were often, like the Diprotodon, of large size when compared with their living allies. Amongst these may be mentioned *Palorchestes azael*, the largest known member of the Kangaroo family, the size of which may be inferred from the fact that its skull measured sixteen inches in length, while that of the largest living kangaroo (*Macropus giganteus*) measures not more than eight inches. Great though its size was, this gigantic kangaroo, judging from the similarity between the bones of its hind legs and those of existing kangaroos, jumped along in leaps and bounds much as its living successors do.

Macropus titan and *M. anak* indicate by the names given to them by Owen their size as compared with living species. Linking the huge Diprotodon with the Wombats was an animal to which the name Nototherium has been given, and which in all probability may be regarded as a gigantic burrowing creature, much like an enormous wombat, though perhaps its size enforced it to be content with digging up roots rather than with actually burrowing amongst them as the living wombat does. There was however an animal, *Phascolomys gigas*, belonging to the same genus as the wombat which reached a height of three feet and was of very massive build.

Fragments of bones show also that at the same time there lived Phalangers (allied to the so called "opossum") of the genus Pseudochirus, and also an animal closely allied to the living "native bear" (Phascolarctos), though in each case the fossil animals were of larger size than the living ones. Another now extinct animal was the Thylacoleo, in regard to the exact nature of which there has been considerable dispute. Lydekker* says " the remarkably trenchant form of the last premolar tooth of this strange extinct representative of the Phalangers not unnaturally led to the conclusion that the creature was a carnivore, preying upon the large herbivorous Marsupials which were its contemporaries, and it accordingly received the specific name which it bears. Fuller acquaintance with its anatomy revealed, however, its intimate kinship with the Phalangers, and when this was fully realised, it was argued that Thylacoleo must be purely a vegetable feeder. Many of the Cuscuses are, however, partly carnivorous in their habits; and in our own opinion it seems probable that in this respect their gigantic cousin resembled them to a certain extent. Not that we mean to assert that Thylacoleo was a creature which preyed on large Mammals, since to attack and overcome such its teeth are clearly not suited; but we do think that it may have probably killed and devoured the smaller Mammals, as well as such birds as it was able to catch."

From its huge size, equal to that of a Rhinoceros, Diprotodon has naturally attracted a large amount of attention, and the recent discovery at Lake Callabonna of a series of complete skeletons will, when the material upon which Dr. Stirling is now engaged has been worked out, enable us to gain a complete account of its real nature. It appears to have been an animal the bulk of which was quite equal to that of the largest Rhinoceros, though it had longer legs than the latter. Unlike a kangaroo both front and hind legs, each of which according to Dr. Stirling were probably provided with five toes, were of the same length, while its tail was only a little over a foot in length. Diprotodon was not therefore a jumping animal, but, like the kangaroo it was a peaceful herbivor, whilst its huge size and strength probably enabled it to tear down at all events the smaller trees, upon the foliage of which it fed.

* Marsupials and Monotremes. Allen's Naturalists' Library, 1894, p. 260.

From Callabonna Dr. Stirling has also reported the occurrence of a fossil of still greater interest in the form of a gigantic bird, some idea of the size of which which may be gained from the fact that its skull is some twelve inches in length and that the length of the hind leg exceeded that of the emu by more than a foot, whilst the whole skeleton is proportionately more massive than that of the latter bird.

Of the fossil remains of extinct marsupials known from the interior of the continent, the most perfect series is undoubtedly that from Lake Callabonna, of which, thanks to the skill and energy of Mr. Zietz, the Assistant Director of the South Australian Museum, a beautiful series has been brought down to Adelaide, where it is now being carefully arranged and set up by Mr. Zietz.

We may now regard as fully established the important conclusion enunciated by the late Mr. Wilkinson, Professor Tate and others that, in times immediately preceding the Pliocene and continuing into the latter, there was a pluvial period during which the now desiccated areas centering in Lake Eyre, Lake Torrens, and Lake Frome, were clothed with a rich vegetation, amongst which lived the large extinct marsupials and birds, the fossil remains of which are found in places such as Lake Callabonna or in the Wellington Caves in New South Wales.

With the gradual desiccation of the interior went hand in hand the extinction of the rich flora and of the large animals dependent upon it, though there was probably some other cause at work aiding in the extinction of the big marsupials, because they became extinct not only in the centre, where the extreme desiccation prevailed, but also in such other parts as Western Victoria, where there is no reason to suppose that the conditions of life, so far as climate was concerned, were rigorous enough to alone account for their dying out. Possibly, as Mr. De Vis has suggested, their extinction was due at all events in large part to some form of senile decay of the race.

Whilst at Henbury a visit was paid by Messrs. Stirling and Watt to the Chandler Range, lying about twelve miles to the north of our camp by the side of the Finke. These hills are composed of Silurian sandstone, and are some 1500 feet high. On them, for the first time during the Expedition, was seen the native fig tree (*Ficus platypoda*), and the pine (*Callitis verrucosa*), both of which were met with after this in abundance on all the ranges formed of Silurian sandstones and quartzites, or of still older gneissic, Pre-Cambrian rocks, as in the main McDonnells.

In the Chandler Range is a curious shallow cave which was investigated by Dr. Stirling and is described by him in the Anthropological section. It has a projecting ledge of rock, and has evidently, for many years past, been associated with some ceremony of the blacks. The front of the ledge, which is about ten feet high, is ornamented with alternate vertical lines of red and yellow ochre, amongst which are dark bands of what is evidently blood which has flowed over the edge and dried up.

The blacks assured us that at this spot a special rite of blood-letting was enacted, in connection with a ceremony, the object of which was to increase the number of wallaby. The native name for the spot is "Antiarra," and it is very probable that here, for many years past, the blood-letting ceremony has been periodically enacted.

Ceremonies of this kind are not uncommon amongst the natives of Central Australia, as will be seen by reference to Mr. Gillen's article in the Anthropological section, where he describes two, one of which is connected with the increase of the "witchetty" and the other with the supply of water. These ceremonies, which may be described as "sacred" corrobborees, are intimately associated with certain subdivisions of the tribe and very clearly indicate the existence of totems—that is, of the intimate connection of a group of individuals with some natural object though this totemistic idea has become considerably modified amongst the Arunta tribe when compared, for example, with those of the Urrapunna tribe to the south of Charlotte Waters and others (as described by Messrs. Howitt and Fison) in which every individual belongs to some particular totem, and in which, further, only individuals belonging to particular totems may intermarry.

In tribes in which the totem subdivisions regulate marriage it is most frequently the case that the individual is supposed not to kill or eat the natural object bearing the name of his totem; but in the Arunta, so far as at present known, no such relation between the individual and his totem is recognised, or if recognised at all then only during perhaps some special period.

A large number of blacks were camped out in the sandy bed of the Finke, amongst the big red gum trees (*E. rostrata*), which here, as elsewhere, grow right in the bed of the creek itself. Some of them were shaping spears, others were grinding Munyeru, but the great majority were lying about doing nothing, and perfectly happy because they had enough to eat—a bullock having been just killed, of which they had, as usual, secured the parts not wanted by the white men.

Whilst collecting various articles amongst them, I was surprised to find that, at first, even the offer of tobacco was firmly refused for a very commonplace necklet, which was, apparently, only a narrow cord of hair well greased and covered with red ochre after the native fashion. After some little time, during which the owner seemed very unwilling to refer to it at all and only spoke in whispers, it appeared that the necklet contained the hair of a dead warrior taken from his head after death. It was therefore regarded really as a charm and as endowing the wearer with the attributes of the dead warrior. It was only after some two hours' persuasion and a liberal gift of tobacco that the owner could be induced to part with it.

Travelling north-west from Henbury we still followed up the Finke. Along the banks were patches of Salsola, but they were becoming less frequent as we left the Desert Sandstone behind us, and grasses such as clumps of Spinifex (*S. paradoxus*), and now and again of Porcupine (Triodia sp.), were becoming more frequent. For the first time also we met with *Eucalyptus gamophylla*, one of the Mallee gums, that is, those which have a bole or bossy stem often not conspicuous above the ground from which arise a number of small branches. The so-called mallee root is, in fact, the main trunk. The leaves of this species as its name indicates have no leaf stalks, but are joined to one another in pairs by their bases uniting around the stem. The tree does not attain to a greater height than perhaps fifteen feet or at most twenty.

All through the scrub we met with large webs of a spider which exists in great numbers throughout the central region from Oodnadatta in the south, up to Alice Springs in the north, and away to the George Gill Range in the west. The webs stretch across from tree to tree for a distance often of twelve to fifteen feet, and are, perhaps, five or six feet in height in the centre. During the day-time the spider (*Nephile eremiana*) can almost always be seen in the centre. It is of considerable size, the whole body being sometimes upwards of two inches long, and when disturbed it rapidly retreats along one of the strong side lines leading away into a shrub on which can often be seen its cocoon attached to the leaves. Very often two webs were found close together—in fact, this was of such frequent occurrence as to draw special attention to it - one with a large and the other with a smaller spider on them but in every case they were all females. Their food consists of any kind of insects which fly into the webs which are so strong as to make riding through them not at all comfortable; they are so strong as to suggest the idea that they might even entangle small birds, but none were seen thus captured. Though the spiders exist in great numbers, still, of the enormous number of young ones contained in each cocoon only comparatively very few can ever reach maturity.

Two days' travelling from Henbury brought us to a spot known as Running Waters, on the Finke. Here the water, which has been following the river bed beneath the surface, rises up, and the rare sight of water actually running for a short distance is seen, but it soon sinks down again and leaves nothing but the dry sandy bed. There must be at this spot some bar of rock over which the water is forced to rise. In the water-hole by which we camped for the night were plenty of fish and an abundant growth of Potamogeton, Vallisneria, and a plant (*Naias major*) with small, thick, succulent leaves. We also secured specimens of the crayfish, which turns out to be the *Astacopsis bicarinatus*, which is widely distributed over Eastern Australia, from Queensland in the north to Victoria in the south, and which is apparently also widely scattered over the interior. The wide distribution of this particular species is evidently associated with its capability of adapting itself to life under varying conditions. In Victoria we always find it in creeks and water-holes; but when the latter dry up it will make a burrow and throw up a cast perhaps a foot high, with a small tubular passage leading through it to the underground chamber, where a supply of water is kept. Under normal conditions in Victoria it remains in the water, and what is called the "land crab," which always lives under logs or in burrows on land, often far away from water, belongs to a different genus (*Engæus*, sp.). This genus is only found in Tasmania and Southern Victoria, and does not extend up into Queensland or even New South Wales. In the more northern parts, as in Queensland, the *Astacopsis bicarinatus* will go on to and burrow in land away from water just like the Engæus, and it is this power of adaptability to varying environment which enables it to survive in Central Australia.

Like the true land crab (*Telphusa transversa*) before referred to, when the drought comes it retires into a burrow in the banks of the water hole. The natives know well and appreciate it as an article of food, for sometimes it grows into the size of a small lobster. They call it illya-aúma.

Another day's journey brought us right into the James Range and into quite a different class of scenery. The monotonous plains, with their alternation of stony and loamy flats, with now and then a patch of sandhills and in the distance flat-topped, terraced hills, were left behind, and we found ourselves travelling through valleys of varying extent, all thickly covered with scrub and lying between rugged ranges of red sandstone hills jutting out into bold, rounded crags.

Leaving the Finke, which was coming down from the north, we struck westwards and camped at the foot of the hills near to the Ilpilla Creek at Illamurta, where is an outlying police camp, placed in this far distant spot principally for the purpose of preventing, if possible, the interference of the blacks with the cattle on the runs.

CHAPTER IV.

The Higher Steppes.

The Southern Part of the James Range and the George Gill and Levi Ranges.

The James Range—The Police Camp at Illamurta—Collecting amongst the Ranges—First appearance of Black Earth—Earthworms—Significance of the presence of Acanthodrilus and Microphyura in Central Australia— The Ilpilla Creek—Persistence of Land Mollusca amongst the Ranges—Fish in the Waterpools in the Ilpilla Gorge—Absence in Central Australia of anything like a great Mountain Range with sheltered and fertile Valleys—Necessity of being in the district during the various Seasons—Leave Illamurta and travel on to the Palmer River—Camp near to the Illara Water hole—Native Tobacco Plant—Absence of Frogs and other animals probably due to low temperature at nights—The Party divides into two sections, one going to Tempe Downs the other to the Petermann Creek—Tempe Downs Station—View from the Station Range—The Walker River and Gorge—The habits of the Porcupine grass Ant—A Corroboree at Tempe Downs—Musical Instruments amongst the Blacks—The Main Camp at the Petermann Creek—Traverse of the Levi Range by Mr. Watt—From the Camp on the Petermann to Trickett Creek and along the southern face of the George Gill Range to Bagot Creek—Our Camps at Bagot and Reedy Creeks—Description of the Reedy Creek Camp—Gum Creeks—View from the Escarpment of the George Gill Range—Collecting amongst the Sandhills to the south of the Range—Jerboa-rats, Mice and Antechinomys—Tracking of Emus by the Blacks—Penny Springs—Cycads, Encephalartos macdonnelli—A Picturesque Gorge—Native Rock Drawings at Reedy Creek—Pigments used by the Natives—Division into Two Parties—The Main Camp travels eastward to Laurie Creek and then to the McDonnell Range—A Small Party under the guidance of Mr. E. C. Cowle goes south across Lake Amadeus to visit Ayers Rock and Mount Olga.

THE James Range is the name given to a large number of ridges which run roughly parallel to one another from east to west. In the Geological section of the work it is used to include the ranges usually spoken of as the James, Krichauff and Waterhouse and also the Silurian or Ordovician Ridge forming part of the McDonnells.

On its way down from the north the main Finke cuts its way across range after range in a series of deep gorges, and everywhere the hills are intersected with valleys of various size hemmed in by more or less precipitous rocks of red sandstone, on which grow fig trees and pines.

The Police Camp lies at the opening of one of these valleys, down which runs a creek, dry at the time of our visit except for one or two small water-holes. A carefully tended garden by the creek side yields an abundant supply of vegetables, and here for the first time we saw what might be called black earth, which was more or less moist and very different from the dry, sandy and loamy ground elsewhere.

At the Police Camp we met Mr. Daer, the officer in charge. Mr. E. C. Cowle, who was associated with him and has now succeeded Mr. Daer as officer in charge, and who we hoped would join the Expedition for a time, was away, having gone over to a place called Erldunda, some distance to the south, in order to meet us, as it had been at first arranged that we should travel by that route. To Mr. Daer we were much indebted for his kindly reception of us and for his generosity in placing horses at our disposal in order to allow some of us to make a flying visit to Ayers Rock and Mount Olga apart from the main camel team, which travelled too slowly to allow of such a detour ; and it is with deep regret that the members of the party who experienced the kindness of Mr. Daer have since heard of his death.

We made our camp at the entrance to the valley and at the base of a picturesque rugged bluff of red sandstone. Spelling for a day, we had the chance of a little quiet work for the first time amongst the ranges, and each of us set out in quest of what he most desired. Messrs. Tate and Watt were out geologising, the former also in search of plants ; Dr. Stirling was busy with the blacks ; Keartland was in search of birds, and amongst the scrub in the gorge behind the camp he secured for the first time a new honey-eater (*Ptilotis keartlandi*), which was afterwards met with in other parts amongst Mulga and Mallee scrub ; Pritchard, one of the two prospectors who had joined us at Henbury and who, together with his partner Russell, was always ready to help us in any possible way in our work, went out with myself to dig along the creek banks. I was anxious to learn something with regard to the earthworm fauna of the centre of the continent, and this was the first spot in which it was at all likely that any such animals could exist. They had previously been searched for, but always unsuccessfully. We began to dig in a patch of damp black earth where reeds were growing thickly. First of all a few specimens of the frogs *Limnodynastes ornatus* and *Hyla rubella* were found, and then a good many snails (*Thersites adcockiana*) were turned out of the ground. After a short time we came upon a small earthworm—the only kind met with during the whole Expedition, though they were carefully searched for.

It is a species of Acanthodrilus (*A. eremius*) peculiar to this part, and belongs to a genus of earthworms which is rare in Australia, where it has only been found in Queensland and in one locality in North west Australia, but which is characteristic of New Zealand and is also found in New Caledonia, Kerguelen, South America and the Cape of Good Hope.

It is a remarkable fact that the earthworm found in the centre of the continent belongs to a small series of animals the ancestors of which may perhaps have reached Australia by way of the Antarctic regions. On the whole the evidence is in favour of regarding this earthworm as a lingering relic of a fauna derived from the north eastern side of the continent during the period prior to the time at which the interior began to assume its present state of desiccation. Side by side with Acanthodrilus lives a small land-shell, *Microphyura hemiclausa*, belonging to a genus which Mr. Hedley has described as being "of high antiquity and of Antarctic origin." This genus also is found in Queensland, New Caledonia and the Loyalty Islands.

Probably Acanthodrilus entered Australia by way of a land tract extending perhaps not at one, but at intervals of, time, from the north of Queensland southwards, towards the Antarctic, across what is now New Zealand, which is the headquarters of the genus. At all events there is no trace of the latter in Tasmania, Victoria or New South Wales. In the centre it is very local indeed, probably it exists elsewhere, but we only found it in three sheltered and favoured spots, separated by long distances from one another. These three spots were Illamurta in the James Range, Bagot Creek on the south side of the George Gill Range and the Finke Gorge in the McDonnell Range. Even in these places it was only found in small colonies each of which occupied not more than a few square yards of ground in extent.

We may, as said above, regard *Microphyura hemiclausa* amongst the molluscs and *Acanthodrilus eremius* amongst the worms as the descendents of ancestral forms which long ago entered the continent, and under very different climatic conditions from those which now prevail, migrated inland from the north-east coast and with change of climate have been left stranded in the interior.

Certainly there cannot have been for a very long time past any introduction of earthworms into the interior, and it is scarcely likely that this genus—the rarest of all earthworms in Australia—should owe its existence in the centre to a chance introduction, whilst none of the, at present abundant and characteristic Australian genera, have been similarly introduced.

Not far from the camp at Illamurta was the rocky ravine through which flows in flood time the Ilpilla Creek. On the rugged sandstone cliffs bounding the ravine, pines and fig trees were growing and under the shelter of the latter various molluscs were found alive. We had before seen plenty of dead shells in the rejectamenta of the Finke, left behind when the floods had dried up, but amongst the Silurian Ranges we now found them alive in their homes from which in flood times they are washed away down stream and are left stranded, when the waters dry up, in places where they cannot live. They prefer the shady sides of the mountain ranges or of the gorges. In some cases, as Professor Tate remarks, one species such as for example, *Thersites adockiana* (which was found in the Ilpilla ravine) will be more or less widely distributed, but in other cases there may be only two or three colonies separated from each other by long distances. It is perhaps a matter of some surprise that so many species of mollusca should be met with in such a district as the dry interior of Australia, but though there is now but little chance of its being stocked from without by carriage of the animals across the dry regions which everywhere separate the central ranges from the moister coastal district, still there is no reason why a considerable number should not have persisted, some in modified form, as the descendents of a once rich molluscan fauna. Water snails such as Limnea, Melania and Bulinus can always find sheltered water-holes amongst the ranges where they can remain alive—if they happen to be developed in them—during even the driest season and from which in flood times other pools can be stocked. Probably like most animals in this district they have acquired the power of reproducing the moment conditions are favourable, and of developing rapidly.

The land forms of snail are more or less hardy, and by means of living as they do in *débris* around the base of and sheltered by thick trees such as the native fig, growing on the shady side of hills, and by plugging up the mouth of the shell to prevent desiccation they can withstand a dry climate in which at first it might be thought that no land mollusc could survive. Naturally most of them are of small size—some exceedingly minute—and not even the largest of them is too big to crawl into a small cleft and so hide itself during the hot season, protected both by its position and its parchment-like operculum. Some of the water forms as previously mentioned (*Isidorella newcombi*) burrow in the earth whilst the hardiness of others (as *Bithinia australis*) is shown by the fact that with their operculum tightly closed they can remain alive—quite dry—for at least fifteen months.

The floor of the gorge was rocky, and in the small water-pools amongst the deep clefts and amidst a rich growth of Vallisneria and Chara were plenty of fish, now so thickly aggregated that they could easily be caught by means of light spears—an art in which the blacks were adepts. Most of them were bony bream (*Chatoessus horni*), but none reached the size of those found in the deeper pools such as the one at Henbury.

Only one example of the rarest of the Central Australian fish *Plotosus argenteus* was found. This is a new species of Siluroid closely allied to the common cat fish (*Copideglanis tandanus*) of the Murray River, though it is much smaller in size, only reaching a length of about five inches. The black boy with me regarded it as a dangerous animal to touch, probably because of its strong dorsal spine.

Except in the case of pools such as the one at Henbury, every individual fish which gets washed down from the small permanent holes amongst the ranges must inevitably perish. It would not be more than at most three or four weeks before all those which we saw along the Hpilla Gorge would be dead, as the pools were very small and shallow and were rapidly drying up.

Leaving Illamurta, we travelled westwards through Mulga scrub along the southern base of the range with its series of jutting promontories. Every now and again were patches of desert oaks or Mallee gum and the hard sandy ground was covered with yellow kangaroo grass, while occasionally there were tussocks of porcupine grass brightened with the red flowers of Brachysema growing around their bases. Though the country with its bold red ranges was somewhat picturesque, at all events in comparison with the monotonous gibber plains, still everything was as dry as in the Desert Sandstone district, and we had now been completely disillusionised with regard to the idea with which we had started—that we should find these central ranges of the continent an oasis in which had been preserved relics elsewhere lost of a more or less primitive fauna and flora.

As Professor Tate has said in the Botanical report, he had "pictured a vast mountain system capable of preserving some remnants of that pristine flora which had existed on this continent in Palæocene times—probably a beech, possibly an oak, elm or sycamore." For my own part I had hoped to find amongst the ranges well watered and fertile valleys, with at all events a few types of animal life, especially amongst marsupials, which had persisted in this isolated part of the continent.

The fact probably is that travellers, struck with the beauty of certain spots, after passing for long, weary weeks or even months over desert country, have unconsciously exaggerated their beauty and fertility. In reality the ranges form bare and often narrow ridges separated from one another by dry and sandy, scrub covered flats varying in breadth from a few hundred yards to many miles, and there is nothing like a great mountain mass with sheltered, well watered and fertile valleys such as we had pictured in imagination.

There are, however, spots such as the Glen of Palms, Ayers Rock, Mount Olga and the gorges amongst the James and McDonnell Ranges, the beauty and even grandeur of which are undeniable, and though the interior did not reveal such forms of striking interest as we had hoped to find still the animal and plant life with its adaptation to a harsh climate was well worth studying.

In the limited time at our disposal we did as much as was possible, but we should have been more contented with a much longer time. With regard to this Mr. Horn most generously made no definite conditions, but as the members of the scientific staff could not possibly remain in the field, owing to University duties, for more than four months it was impossible for us to do more than was done. In Central Australia much depends upon conditions of climate. Especially, so far as the fauna is concerned, you need to be there in a wet season as well as during a dry one. Just before and after the rains animals are in evidence which are not seen at other times, and of course within a short time of the rainfall plants spring up and blossom which are never seen in the dry season.

Along with the plants go the insects and along with these to a large extent the birds, reptiles, amphibians and smaller mammals. We were not fortunate enough to meet with any rain and our collection suffered in consequence, but thanks to Mr. Byrne of Charlotte Waters, Mr. Gillen and Mr. J. Field of Alice Springs, Mr. Cowle of Illamurta, and to a second visit after the following rainy season I have been largely able to supplement the zoological collection. Much doubtless yet remains to be done, but both as regards Botany and Zoology the important features are now probably known.

Some twenty miles west of Illamurta we struck the Palmer River close to the Illara water-hole—a deep pool fringed with rushes and hemmed in with high rocks, in the shelter of which close to the pool was a fine growth of the native tobacco plant (*Nicotianum suaveolens*). At the edge of the water was a dense bed of the curious water plant *Naias major*, some twenty feet in length, six feet in width, and in the thickest part four feet in height, but unfortunately no fruit was to be seen. A few wood duck were swimming about, and small shoals of *Chatoessus horni* and *Therapon percoides* were feeding, but were too wary to be caught. To the south of the river was a wide open flat, but the northern bank for two miles was hemmed in by an escarpment of rock at the base of which the river ran, that is to say it would run in flood time, after emerging from the rocky gorge at the mouth of which lay the Illara water-hole. The bed was filled with rushes and contained a series of shallow water pools in which I expected to find numbers of frogs, but was disappointed, as not one was to be seen or heard.

This was perhaps due to the fact that as soon as ever the sun set it became very cold, the temperature gradually falling to several degrees below freezing point. It was now the first of June, so that we were very close to mid-winter, and what frogs there were remained hidden, or at all events kept away from water during the cold nights.

After crossing the Palmer we divided next morning into two parties; the main camel train, with Messrs. Winnecke and Watt, went south-west into the Levi Range, while Professor Tate, Dr Stirling and myself followed up the Walker River, which here joined the Palmer, to Tempe Downs—a cattle station then in the possession of Mr. F. Thornton, by whom we were most kindly received and entertained. Tempe Downs is the most westerly of the few stations or, to speak more correctly, cattle runs in the central district; of late years drought and low prices have combined to render the enterprise of those who have attempted to utilise the land of the far interior a somewhat hazardous undertaking. The outlying runs are managed by one or two white men aided by black "boys." Occasionally there has been trouble with the natives, to whom, in hard times, the sight of cattle must be a great temptation; but by kindly treatment of them Mr. Thornton has had comparatively little trouble with the aborigines. It is not difficult to realise that it must appear exceedingly strange to the blacks that whilst the white man can shoot down the emus and kangaroos he, the blackfellow, is not allowed to spear the cattle.

Tempe Downs is situated in a long valley (in fact it is, strictly speaking, the valley to which the name given by Giles applies) not more than a mile in width, open towards the west end but completely closed eastwards except just at one spot where the Walker River, which runs along the valley, breaks in a deep gorge through the mountain range, closing in the valley on the north. By running a fence across the valley to the west and another across the narrow gorge of the Walker there is formed a paddock many miles in length in which, if need be, all the cattle can be kept securely.

From the Station Range on the south, which rises five hundred feet above the valley, which is itself 1670 feet above sea level, we had a fine and characteristic view.

To the north of us was range after range of hills running east and west and separated from one another by a series of parallel valleys; a mile to the east the Walker broke through the nearest range to continue its easterly course to the Palmer in the valley just beyond. The hill on which we stood was cut through by deep gorges, in many of which were little rock-pools, whilst the valley at our feet stretched away westwards for twenty miles to broaden out into a plain lying to the north of the George Gill Range and opening into the desert sandhill country out to the west of the main mountain ranges.

We spent nearly three days at Tempe Downs, collecting amongst the valleys and on the hill sides and photographing amongst the blacks, a good number of whom were camped out in the bed of the river close by. Everything was as dry as usual except for a fair sized water-hole in the shelter of the Walker Gorge, in which we secured a new species of fish (*Nematocentris tatei*), which we afterwards found in other water-holes amongst the ranges. Mr. Thornton told us that after a heavy fall of rain small fish are often found in the pools amongst the ravines high up on the hill sides. The only possible way in which they can get into such positions is by being carried there perhaps in the form of eggs attached to the feet of birds, which the moment the rain comes, appear as if by magic.

In the same water-holes were abundant specimens of a mollusc, *Melania balonnensis*, crawling upon the floor, while, as usual, the sandy edge was marked with the ridges left by the mole cricket (*Gryllotalpa coarctata*) as it burrowed its way along the damp sand. We could always secure this insect at the edge of water-holes with a sandy margin.

On the rising ground between the Walker and the Station Range, in a scrub of gums, Acacias and Cassias, were somewhat open patches covered with tussocks of porcupine grass (Triodia sp.), and here I spent some time watching the habits of a curious little black ant which had been described by Mr. Kirby under the name of *Hypoclinea flavipes*, and may be called the Porcupine grass Ant.

Various explorers have already noted the presence of curious little galleries which run along the surface of the ground, often for long distances, from one tussock of porcupine grass to another. In some parts, and especially on hard sandy soil where the tussocks of grass are not too close together, these galleries as they are called, though tunnels would be a better name, form a regular network. Each is from one quarter to half an inch in width and, in section, is roughly semicircular in shape. They are made of grains of sand fastened together with the resinous material obtained by the ants from the gelatinous leaf sheaths of the Triodia, and they form runs which lead from one tussock to another, along which the ants can travel sheltered from the light and more especially perhaps from the heat of the sun. In many cases they lead for long distances up the trunks of gum trees.

In addition to the galleries running along the ground some of the tussocks of Porcupine have their long spiny leaves more or less wholly enclosed in little

cylinders of sand, formed in the same way, to such an extent that the whole tussock looks like a network of sand tubes. In other cases there were only small cylindrical cases of sand here and there on the spiny leaves. Each of these was perhaps half an inch or an inch long and a quarter of an inch in diameter, and so built that the grass blade formed one part of the wall, a space being enclosed between it and the sand. The cylinder was always closed at the top and had a small opening at the bottom, so that if rain came it would not get into the chambers.

Watching the ants, which are very small and black-bodied with yellowish feet, I saw them constantly running in and out of these chambers, and on opening the latter found that they were always built over two or more Coccidæ attached to the leaf of the grass. Here, as in the case of the ants described by Belt in Nicaragua, the Coccidæ abstract nutriment from the leaf, and the ants take advantage of the exudation from the body of the Coccus. This arrangement is without doubt of advantage to both parties concerned. The Coccidæ gain protection from enemies, to whom they are made invisible, and also from the great heat of the sun, and at the same time the ants get without much trouble to themselves a supply of food.

I think after examining a considerable number of tussocks of porcupine grass both here and elsewhere that the network of sand tubes, which as above said sometimes cover the whole tussock, always commences in the form of a number of chambers specially built over the Coccidæ which it is quite likely — though I had not the means of testing it — are actually brought on to the leaf by the ants. Then covered passages are made up the leaves, leading from one chamber to another and so gradually the whole tussock is enclosed.

Tracing the passages down to the roots the ant nest is seen to be built around the latter or rather part of it as the tussocks are often of large size. The nest consists of a more or less conical mass of material built up of sand particles agglutinated by the resin derived from the leaf sheaths with remains of the roots running through it. The largest nest dug up measured a foot and a half in depth and about a foot across at the top which was just below the surface of the ground from which it gradually tapered downwards. It was riddled with passages of various sizes, some an inch in diameter, along which the eggs appeared to be lying about irregularly. Each nest contains larger and smaller winged forms, small black and larger brown-black wingless ones.

The blacks gather together large quantities of the viscid leaf sheaths of the porcupine grass (*Triodia pungens*), and after cutting it up into small pieces, burn away as much as possible of the grass itself and so obtain lumps of black resin in which remnants of the leaves and leaf stalks can always be seen. These lumps of resin they use, after softening with heat, for various purposes such as that of fastening bits of flint on to the ends of their spear-throwers, the resin setting into a mass as hard as stone.

In an appendix to the Botanical report Mr. Maiden has shown that the nest is built of sand particles agglutinated with resin, a coating of ferric oxide giving to the whole the colour and appearance of reddish-brown clinker. That the ants do carry away the resin from the leaf sheaths of the grass, both to make their nests and to cement together the sand particles of which their tunnels are made, cannot be doubted ; but it is rather difficult to understand how they accomplish the task, as the resin on the leaf sheath is of such a nature that it feels like varnish, and it would rather be thought that the ants would have stuck to the resin when they touched it and would not have been able to carry it away in a condition fit to use it in the way in which they do. Perhaps they have the power of smearing some fluid matter over it which enables them to carry away little pieces of it without its adhering closely to their appendages.

In many cases the nest was at the base of a tussock which had evidently been burned.

The blacks—so they assured me—do not make use of the resin already massed together by the ants to form their nests, but always get it by burning the leaves of the porcupine grass for themselves. They could not tell me why they did not do so, but only said that their fathers had not and they never did, which is a typical answer to the question "why" when put to a black. This was at Tempe Downs, perhaps in some other parts of the continent they do make use of the ant nest.

Mr. Maiden has stated that they do so in certain parts of West Australia, and certainly the first time you pick up a piece of the nest you are likely to jump to the conclusion that it is exactly what they would do, as at first sight the clinker-like mass resembles the blocks of resin which the blacks use for the purpose. It seemed such a foregone conclusion that when the black boys told me they never used the resin from the root I was at first considerably surprised. However, on second thoughts, it did not appear so surprising. The main part of the nest consists of sand grains, and without burning it completely away it would be no easy task to separate the resin from the sand, whilst in the case of the leaves of the plants it is easy to burn them away and leave the melted resinous mass behind. I have not, amongst a large number of native implements examined, seen one in which sand grains, such as would certainly be found were the resin obtained from the ant nest, could be detected; but, on the other hand, it is rarely that little bits of the leaves of the grass cannot be seen which have escaped the fire.

Amongst the roots of the grass I also found a larger form of Coccus and also one special form of bug, but could not detect any special connection between them and the ants.

Our stay at Tempe Downs gave Professor Tate the opportunity of examining some fossiliferous beds from which Mr. Thornton had previously obtained remains of Trilobites, and the fauna of which was now more fully worked out.

Amongst the blacks a considerable collection of native articles of various kinds was made by Dr. Stirling, and as this was the most westerly spot at which we came in contact with them in any number, and as men both of the Luritcha and Arunta Tribes were gathered together, Dr. Stirling's time was fully occupied. At night time corrobborees were held. A place was cleared in the scrub and fires lighted at either end. At one end sat the audience, whilst the performers danced up and down the open space keeping time to the chanting of the audience, who also beat upon the ground with sticks. The fitful light shining on the white trunks of the gum trees and on the decorated bodies of the performers and the low monotonous chanting, at one time breaking into a louder refrain and then dying away into a murmur, produced a curiously weird effect. Each corrobboree has its set parts or "figures," and each performer has his own part to play, for of course the dancing is confined to the men, the women being merely spectators. Each corrobboree also is associated with some special object such as, for example, the emu, or wild cat (Dasyurus) or wild dog (Dingo). The one which we saw at Tempe Downs has been described by Dr. Stirling in the Anthropological section.

Whilst the chanting is not by any means devoid of a curious and quaint musical rhythm of a simple nature, the musical instruments are of the simplest and most primitive nature.

In the first place, if it can be called a musical instrument at all, there is the flattened stick with which, by beating monotonously on the ground, the time is emphasised. In other parts of Australia a dull, heavy sound is produced by beating upon rolled up fur rugs, but the Arunta and Luritcha blacks do not manufacture anything like a rug.

By way of trumpet there is a hollowed out piece of the stem of a tree perhaps two or three feet in length and ornamented externally with red ochre and bands of white gypsum and yellow ochre. This is blown through to increase the volume of sound.

The only other musical instrument which the Central Australian blacks use is one, the native name of which is "Trora," which was given to me by Mr. Byrne at Charlotte Waters. It has simply the form of two pieces of wood, each of which is about six inches in length rounded off and tapering at either end. One, which is somewhat the larger, being about four and a half inches in length and an inch and a half in diameter, is held in the left hand and is struck at intervals by the other held in the right hand. The latter one may be varied in form and sometimes has the terminal part in the shape of two prongs. Beyond these the Central Australian blacks do not seem to have any musical instruments. They have not conceived any idea even of a drum.

Whilst we were busy at Tempe Downs the camel train, with Messrs. Winnecke and Watt, had camped in the valley to the north of the Levi Range, and this spell gave to the latter the opportunity of traversing and examining carefully the series of ridges which form the range. Though not of any great height, still their rugged nature and the steep faces make climbing somewhat tedious work. Mr. Watt found that the Levi Range has been worn out of a gentle syncline; on both the north and south side is a bold precipice from two to three hundred feet in height, below which the sides slope down for some two hundred feet more at a steep angle, covered with a talus of blocks of various sizes and overgrown with thin scrub, above which the red escarpment stands out boldly.

The main camp to the north of the range was by the side of Petermann Creek where it sweeps round in a big curve hemmed in by a great amphitheatre of bold red cliffs.

Following up the Petermann for some miles we came to the western termination of the Levi Range and turned southwards between the latter and the George Gill Range. If it were not for this break, through which flows Trickett's Creek, the George Gill and Levi Ranges would be continuous. After coming out from this gap we turned eastwards and skirted the southern escarpment of the George Gill Range.

Away to the south was the desert sandhill country which forms the basin of the Lake Amadeus drainage system, so that, strictly speaking, we had now passed out of the Larapintine region and were on the northern limit of the Amadean from which stretches out westwards the true Desert or Eremian region, of which the basin of Lake Amadeus may be regarded as the eastern termination. The George Gill Range on its southern face presents a succession of bold headlands separated from one another by gorges down each of which runs, in wet seasons, a stream, the more important of which are Bagot, Stokes, Reedy, Penny and King Creeks. Each of these runs out for a short distance—perhaps ten miles at most—into the sandhill country, where their waters rapidly disappear. They give rise to what have always been termed by the early explorers "Gum creeks," that is sandy beds which only contain water, if at all, at rare intervals, but along the sides of which grow a line of gum trees (*Eucalyptus rostrata*).

Our first camp was at Bagot Creek. Here there were two small water-holes, one of them surrounded with a rich growth of the reed *Arundo phragmites* in full flower while a very small stream trickled at intervals down the rocky valley leading up behind our camp into the hills.

In certain respects our camp here and the next one at Reedy Creek were not only amongst the most pleasant from their picturesque surroundings but were the best from a collecting point of view. On the one side we had the range of Silurian sandstone hills with its sheltered water-pools and on the other the open sandhill country. There was plenty of work for the Zoologist and Botanist; within accessible distance were fossiliferous strata, and the presence of aborigines was made evident by rock paintings. A lengthy stay in this part with a main camp by one of the water-holes and time to go out and explore the district would probably yield valuable results.

A description of Reedy Creek will serve to give some idea of the surroundings. Here our camp was at the base of a semicircular hollow in the range open to the south and shut in to the north by precipitous cliffs of red sandstone some two hundred feet in height. At one spot at the base of these sheltered by the rocks and hidden by a growth of rushes and ferns was a deep water-pool. On the rocks were pines, fig trees and Tecomas, and close by the water's edge were clusters of the ferns Adiantum and Cheilanthes, while a rich growth of Aspidium had spread over the swampy ground which formed an outlet from the pool in flood time. The water was flecked with the floating leaves of Vallisneria and Potamogeton, on the stems and submerged parts of which a black water-planarian—the only one met with during the Expedition—was crawling about. Large Nepa-like insects, two inches in length, Notonectas, water beetles, minute cyprids and molluses such as Ancylus and Bulinus made up the water fauna; there were no fish in any of the pools to the south of the range nor were there any frogs to be seen or heard.

From the water-hole the rocks rose with precipitous sides over which at one point there was evidently in rainy seasons a waterfall coming down from a rocky gorge above. Outside the hollow in which was our camp the southern face of the hills though steep could be climbed and the gorge above the water-pool was found to be occupied by a succession of small holes surrounded by bare rock. Evidently during the rainy season these upper pools are scoured out and so they contained very little in the way of animal life. Amongst the rocks the usual pines and fig trees were growing, and a few plants of *Hibbertia glabberima* with its striking yellow flower, the largest of the genus. On the whole range grew plenty of porcupine grass, in fact we found this ranging from the sandhills right to the very summit of the highest mountains. In the gorge above the lower pool were numbers of pot-holes evidently worn out by the grinding action of the stones swept into them when the floods came down from the hills. In one of them was a snake (*Pseudonaja affinis*).

From the top of the escarpment we could see the range running away east and west with its series of bold bluffs rising one behind the other. Westward it terminated in a high scarped hill called by Giles, Carmichael Crag. Out to the south stretched the sandhills, with Mulga and Mallee scrub on the flats, and here and there a low ridge of Silurian sandstone standing out above the surrounding country; while the creeks could be traced running away from the range one after the other with their fringe of gum trees dying away in the distance.

On the hard sandy flats skirting the range we found an abundance of mice and jerboa-rats. Each of them makes a hole in the ground just big enough to admit the body and from this the burrow goes down for perhaps three or four feet. In the mice burrows (*Mus gouldi*) were more than one adult with young ones, usually four in number. In the jerboa burrows (*Hapalotis mitchelli*) there was never more than one adult, with sometimes two broods of young ones also usually four in number.

The jerboa-rat, as is well known, has developed curiously long hind legs just like many of the marsupials, in fact when they are travelling it is not easy to distinguish in colour, size and mode of progression a Hapalotis from an Ante- chinomys. Both live in the same class of country—hard sandy ground covered with tussocks of grass and scrub composed of Mulgas, Cassias and Acacias of various kinds.

The little mice, which live side by side with the Hapalotis, thrive just as well as they do, though they have not taken on the curious mode of progression adopted by the latter and can apparently cover the ground just as rapidly as the jerboa rat can. As pointed out elsewhere (Part II., Zoology, p. 11), the advantage of this mode of travelling would appear to consist not so much in the greater speed attainable as in the greater difficulty which their enemies — the birds of prey — probably find in pouncing down upon and seizing a small animal progressing by leaps and bounds. The Antechinomys is mainly an insectivorous and the Hapalotis an herbivorous animal, but whilst the latter is to be obtained in hundreds it is only very rarely indeed that the former is secured, and in fact, though searching in favourable country, we only obtained two specimens during the Expedition. Though far from common, it has however a considerable range, as specimens have been secured at Charlotte Waters, Hermannsburg and Alice Springs, and it doubtless exists in very small numbers all through the hard, sandy, scrub covered flats of the interior.

Whilst at Reedy Creek I had a good opportunity of witnessing the tracking powers of the blacks. I was out in the scrub with three of them when suddenly they came to a standstill and after carefully examining the hard ground they became very excited. On asking what was the matter they told me that there was an emu about with six young ones. The three then separated and commenced to track it up. They went on a trot the whole time; not a word was spoken but where the scrub was thin they communicated with each other by signs. After two miles' run, during which it was quite enough for me to do to keep up with them and to look after my collecting material without troubling to look after tracks which I could not detect, they came to a sudden halt, and there in an open patch in front of us was the mother emu with its six young ones. The mother at once made off, but, shouting and laughing, the blacks soon caught the young ones and we brought them back to camp and carried them alive for some hundreds of miles on camel back. The ground was so hard that only an experienced white man would have detected the tracks of the old bird, but it did not take the blacks more than a minute's careful examination of the very faint tracks to come to the conclusion as to the correct number of young ones. If they had had their spears with them the old bird would certainly have been captured. Their keenness and suppressed excitement when on the track were worth seeing, as well as their childish glee when they were successful.

A little to the west of Reedy Creek was another gorge amongst the hills out from which flowed Penny Creek. On the rocks enclosing it were growing at one spot a considerable number of Cycads in fruit (*Encephalartos Macdonnelli*).

CYCADS (*Encephalartos Macdonnelli*).

The species is confined to the Higher Steppes of the central region and this was the first occasion on which we had seen it. Growing right on the face of the rock, where it would scarcely be thought that there was earth enough to afford sustenance for so large a plant, they look very picturesque. The older ones have a stem some three or four feet high from the top of which springs a crown of dark green, graceful, palm-like fronds, each of which may be as much as ten feet long.

The gorge led away back into the range, and climbing over the rocks we made our way, disturbing several rock wallabies (*Petrogale lateralis*) as we did so, along a narrow cleft not more than a yard wide and in parts fully one hundred feet in depth, and then clambering down the steep face of a cliff found ourselves in an upper part of the gorge, where the rocks in colour and weathering mimicked on a small scale the cañons of the Colorado district.

In rainy seasons the water must pour in torrents down the narrow bed of this upper part of the gorge, but now there were only small pools amongst the rocks the sides of which lower down, where the valley broadened out somewhat, were thick with rushes and Aspidium. One or two new species of Molluscs were secured and also a curious Orthopteran insect resembling a small flattened-out cockroach which clung almost as closely to the surface of submerged leaves as a limpet to a rock. It was evidently mature and adapted to life in the water, and we only came across it in this one small pool.

Unfortunately we could only spend an hour or two in this spot, which would well repay a stay of several days, and if we were to reach and have any time in the main McDonnell Range it was essential for us to lose as little time as possible, more especially as it had been decided that a small section of the party was to make a flying visit to Ayers Rock lying away to the south across the sandhill country. Had we then known what the main McDonnell Range was like there is no doubt but that we should have lingered longer amongst the valleys and by the creeks on the south side of the George Gill Range.

Close by our camp at Reedy Creek the natives had been ornamenting the rocks with drawings, the most elaborate of which was supposed to represent a view seen from beneath of an Emu sitting on eggs with the characteristic, conventional, three-pronged markings representing the emu tracks leading up to it.

In making their drawings the blacks seem to usually use four colours—black (charcoal), red and yellow ochre and white gypsum. A fifth colour—pink—is sometimes obtained by mixing the red ochre with gypsum. Somewhere to the south of the Levi Range is a patch of red ochre, which amongst the natives is a valuable asset and is traded over considerable distances.

On 16th June we divided into two parties. The main camp with all the camels and stores, accompanied by Dr. Stirling, Professor Tate and Mr. Winnecke, went westwards along the base of the Range. Their intention was to go first to Laurie Creek, lying out to the west of Carmichael Crag, and then to push northwards by way of Glen Edith across the sandhill country to the western end of the main McDonnells. Then following these eastwards the whole party was to meet again in about a fortnight at the deserted Glen Helen Station close to the base of Mounts Zeil and Sonder. The second party, on horseback and equipped as lightly as possible, was to go south under the leadership of Mr. E. C. Cowle across Lake Amadeus and then visit and photograph Ayers Rock and if possible Mount Olga. This party consisted of Mr. Cowle, the leader, and Messrs. Watt, Belt and myself, together with one of Mr. Cowle's black police trackers, Larry by name.

I gladly take this opportunity of expressing our appreciation of the cordial manner in which Mr. Cowle fell in with our plans and aided us in our work which must at times have appeared somewhat strange but which could not have been carried out except by his help. I am personally, as will be seen by reference to the Zoological report, very much indebted to Mr. Cowle's exertions in procuring specimens of interesting animals such as the honey ants, full series of which we had not the opportunity of securing during the Expedition. It may also be added here that throughout the Expedition we received the most ungrudging and valuable help from all with whom we came in contact.

CHAPTER V

The Desert Country.

From the George Gill Range to Ayers Rock and Mount Olga.

Our Equipment—Photographing in Central Australia—Departure from Reedy Creek - Camp for the Night after travelling sixteen miles—Sandhill Gum Trees—Winnall's Ridge, the most Southern Outcrop seen of Silurian Quartzite—The Pituri Plant—Uses to which it is put by the Blacks—Kangaran's Well—A most unlikely Spot for Water—The Remains of a broken down Mound Spring—Dingoes in the Water—Reach Lake Amadeus at Sunset—Cross the Salt Bed and Camp on the South Side—The Present State of Desiccation of the Lake Amadeus Area—Leave Lake Amadeus—Coulthard's Well—Travel all Day over Porcupine Sandhills and in the Afternoon Reach Ayers Rock—View of the Rock from the Sandhills—Camp by a Small Water-hole in a Chasm in the Rock—No Permanent Water at Ayers Rock—Spend the Day round the Rock—Native Drawings on the Walls of Small Caves—Honey Ants—Tadpoles of Heleioporus pictus in the Water-hole—View across the Plains towards Mount Olga at Sunset—A Family of Sandhill Blacks—Ride across the Plain to Mount Olga—Camp at the Entrance to a Deep Ravine—Tietken's Marked Trees—No Permanent Water at Mount Olga; only a Small Rock-Pool now remaining—Camp at Wild Blacks—Ride back to Ayers Rock—Cooking of a Kangaroo by the Blacks—Return to the George Gill Range—Increase of the Water in Bagot Creek—Crossing the George Gill Range—Petermann Pound—Cross the Station Range and reach Tempe Downs—Leave Tempe Downs and follow the Walker back to the Palmer—The Gorges along the Palmer—Low Temperature at Night Time—A Large Tussock of Porcupine Grass—Follow the Palmer up to the Missionary Plain and Camp close to Pine Point—A New Species of Grass Tree—Sporadic Distribution of Certain Species of Plants—The Missionary Plains—Glosse Range—Rock Pigeons—Camp in the Southern McDonnell Hills—In the Morning join the Main Party at the Old Glen Helen Station.

WE had with us in addition to our riding horses two pack horses carrying our provisions, a small supply of water and not least in importance the camera, the careful packing of which, to prevent its being completely smashed up as the horses jogged on or sometimes crashed through the Mulga scrub, was not an easy matter. Photographing in Central Australia when on the march from day to day is not altogether pleasant or easy. The light is intense and extra precautions must be taken to prevent light fogging of the plates; but worse than this is the dry heat which was so great during the day time—though at nights the temperature was below zero—that it requires a specially well made camera to stand the combined strain of continuous knocking about and the great range in temperature. I had two cameras with me—a whole and a quarter plate, the latter fitted with an Eastman roll holder. Both cameras were made by Messrs. Watson and Sons, and each of them had before this been taken over some of the roughest parts of Victoria and Tasmania. They were not specially chosen for the work, than which nothing could have afforded a more severe test; but, except for external disfigurement owing to their being thrown off the back of a camel, they suffered little damage and served their purpose admirably.

The jogging of the horses and camels is very liable to smash plates—I lost nearly two dozen of mine in this way—and fine sand grains penetrate everything and often scratch the film. It is almost impossible to avoid this in Central Australia, as they get into the dark slide and upon the plates when you are changing them. To avoid this as far as possible I adopted the plan suggested to me by Mr. Horn of putting over the interior of the dark slides a coat of vaseline, to which the sand grains adhered.

After giving the horses a last drink at Reedy Creek as it was by no means sure when they would get their next, we started out southwards soon after midday. Our track lay across Porcupine-covered sandhills with intervening flats covered with Mulga and Desert Oaks. Every now and again were low-lying and scrub-covered sandstone rises, and away to the north we could see the high ranges stretching east and west. Out beyond Carmichael Crag was a big smoke made by the main party.

After travelling some sixteen miles we crossed the end of King Creek, where the dry bed becomes lost amongst the sandhills and the line of Red gums disappears, and shortly after sundown camped for the night on a hard, dry claypan.

We had breakfast before daylight and just at sunrise the black boy brought in the horses and we started off. All the morning we were traversing low sand-hills, on many of which grew a fine sandhill gum, *E. eudesmoides*, which reached a height of 50 to 80 feet. The trunk is silver-grey in colour and very shiny, except the butt where it is covered with a paper-like bark which peels off in long, yellow-brown scales. The grey-green foliage usually forms a kind of umbrella shaped mass, and it is somewhat strange to find a big tree like this right out amongst the waterless sandhills.

About twelve miles south of our camp we passed by the eastern end of Winnall's Ridge, which forms a narrow Silurian quartzite hill rising abruptly, with a well-marked escarpment, some three miles long on its north side. It forms the most southern outcrop of Silurian quartzite which we came across in our journey. The height of the ridge is about 1700 feet above sea level, and that of the escarpment at its eastern end about 200 feet. On the sandhills round about, the Pituri plant, *Duboisia Hopwoodii*, was growing. It forms a small, stiff shrub usually about four or five feet high with coriaceous, lanceolate leaves which are used in some parts by the blacks as a narcotic, and as an article of trade it has considerable value amongst them. In this part they seem to prefer the native tobacco plant, *Nicotianum suaveolens*. They either simply roll a few of the leaves together and then suck or chew them, or else cut the leaves up finely and mix them with ash obtained from burning the leaves and twigs of a bush, preferably a Cassia. The leaves and ash are made up into little plugs, which are held when sucked so as just to protrude through the lips. The chewed mass when not in use is tucked in safely amongst the well-greased ringlets. If you put your hand up to your mouth and pretend to suck something a black fellow will at once know what you mean, and will in all friendliness offer you his well-used packet of tobacco leaves or his "plug" for a "chew."

The chief use of the Pituri plant in this neighbourhood (apart from its value as an article of barter) seems to be that of making a decoction for the purpose of stupefying and then catching the emu. The leaves are pounded in water and the decoction is placed in a wooden vessel where the emu is likely to come across it, or else a small pool or a fenced-off portion of a larger one is used for the purpose. After drinking it up the animal becomes so stupefied that it falls an easy victim to the blackfellow's spear.

Just to the south of Winnall's Ridge lies a small flat in which, surrounded by tea-tree, is a small native well, known to the white man as Kamaran's Well and to the blacks as Unterpătă. The accidental discovery of this small water supply by Kamaran, one of Gosse's men, was the means of enabling the latter to cross this otherwise waterless track. It lies right amongst the sandhills where the existence of a spring would never be expected.

The well is evidently the remains of a broken down mound spring and has the form of a hole some fourteen feet deep and perhaps ten feet across at the top, the walls slanting steeply down until at the bottom, where lies a pool of water, it is not more than four or five feet across. It is formed in a deposit of Travertine, the remnant of what was once a mound with a spring at the top; the gradual desiccation of the Amadeus basin has gone on until now the underground supply is so small that the water in the spring does not reach the surface. With continued desiccation the water will gradually disappear altogether.

We approached it in the hope of finding a supply for ourselves and still more one for the horses, but to the disappointment of man and beast alike we found it simply stinking with the bodies of five dead dingoes who had ventured into it in search of water and had evidently been too weak to clamber out. All that we could do was to drag out the decomposing carcases in the hope that it would be a little better on our return.

Travelling on after a short halt we came just at dusk to the top of a sandhill and saw Lake Amadeus lying at our feet. It was a strange sight; the bed of the lake was here only some three-quarters of a mile wide, but east and west it stretched away to the horizon, widening out, especially westwards, into a vast sheet many miles across. There was not a speck of water, only a dead level surface of white salt standing out against the rich after glow in the west and the dull sky to the east, whilst north and south it was hemmed in by low hills covered with dark scrub.

It was at this spot that the lake had first been crossed by Gosse and shortly afterwards by Giles, the latter having been previously baffled in his attempts to cross owing to the boggy nature of the ground. By good luck, as Mr. Cowle who had previously been across, was aware the bed was dry and passable, and dismounting we led our horses over with little trouble, and just as it grew dark camped on the top of a low rise on the south side.

Everything was perfectly silent; there was no sign of animal life except for a solitary gaunt-looking dingo which followed us half-way across, and the white sheet of salt seen in the darkness through the sharp, thin stems of the Mulga looked strangely weird.

One could not help thinking of the contrast between the silence and sterility of the scene as we looked down upon it now and the fertility and abundance of life which must once have characterised it when in bygone ages it was a great sheet of fresh water surrounded with a rich and varied forest growth amongst which browsed huge diprotodons and birds as large as the New Zealand Moa.

The day had been hot and somewhat fatiguing, and as this was the second night out for the horses without water they had to be tied up to prevent them from wandering far away in search of food and drink, as there was another still harder day's work in store for them before, as we hoped, they would get water at Ayers Rock.

After breakfasting by starlight we left the lake and riding through the scrub, in which we passed a mound-bird's nest (*Leipoa ocellata*), came after some ten miles to another native well called by the blacks Kurtitina. This, just like Unterpata, is a hole in Travertine. It is however much smaller than the latter—just large enough to comfortably allow of a man getting down. The main hole curves somewhat and then at a depth of ten feet there lies to one side a smaller hole running down for two feet more in which was a little damp black mud. This was scooped out in the hope that a little water might trickle in before our return.

Riding on all day long we kept mounting one sandhill after another, all covered with tussocks of Porcupine grass, amongst which the kangaroo-rats, *Bettongia lesueuri*, kept dodging in and out with remarkable speed and agility.

Whilst we were riding along in this part of the country our attention was drawn by Mr. Cowle to a small rat-like creature which was running about, and dismounting we captured it after a smart chase, during which it ran across from tussock to tussock. It turned out to be one of the most interesting of the new animals found during the Expedition. It is a new species of the genus Sminthopsis, which includes the pouched mice, most of which are ground animals, in contrast to those of the closely allied genus Phascologale which are usually described as being arboreal in their habits. In reality this is only partly true as there are species of Phascologale such as the crest-tailed Phascologale (*P. cristicauda*) and the fat-tailed pouched mouse (*P. macdonnellensis*), which are undoubtedly fossorial in habit; as a general rule also the species of Phascologale in addition to having somewhat more stoutly built feet than those of Sminthopsis, have a number of striated pads on the sole which are doubtless of use to them in climbing. The little animal now captured for the first time has from its living amongst the sandhills been called *Sminthopsis psammophilus*. It must evidently be able to exist without any supply of water other than what it gets either from the moisture in its food or else perhaps from the heavy dews which fall during certain seasons of the year, and it was the only small marsupial which we saw running about during the day time, for most of them are strictly nocturnal.

Between the sandhills, some of which were a hundred feet high whilst all ran in a general north-east and south-west direction, were small flats covered with funereal-looking Desert oaks, and where the harder surface of the ground afforded some little relief to the horses, whose feet and legs were tired and sore with toiling over the heavy sandhills on which the Porcupine could not be avoided.

This Porcupine grass, which is often incorrectly spoken of as "Spinifex" in the writings of many of the explorers of Central Australia is one of the most serious obstacles met with in travelling across the desert region of the southern part of the interior. Each tussock when young resembles more than anything else a gigantic pincushion with the pins represented by long knitting needles radiating in all directions. As the tussock grows older and increases in size the inner parts die away leaving a circular rim the diameter of which may be as much as nine or twelve feet. The young leaves are flat, but as they gradually dry each rolls up into a stiff, needle-like cylinder. In one species (*Triodia pungens*) they are covered, as described before, with a very sticky varnish. Not only do these tussocks of

Ayers Rock

Porcupine grass grow so closely together that it is impossible for horses or camels travelling through them to avoid having their legs severely irritated by the pointed leaves, but the sandy tracts which they inhabit are destitute of water.

PORCUPINE GRASS *(Triodia pungens).*

It was with no little relief and pleasure, that after traversing more than thirty miles of sandhills since leaving Lake Amadeus in the morning, we reached the top of the last one and saw the Rock not far away.

Ayers Rock is probably one of the most striking objects in Central Australia. From where we stood the level scrub stretched away monotonously east, west and south to the horizon. Above the yellow sand and dull green Mulga rose the Rock—a huge dome-shaped monolith, brilliant venetian red in colour. A mile in length, with its sides rising precipitously to a height of eleven hundred feet above the plain,[*] it stands out in lonely grandeur against the clear sky. Its otherwise smooth sides are furrowed by deep lines of rounded holes rising in tiers one above the other and looking as if they had been hollowed out by a series of great cascades down which for many centuries the water in the rain seasons must have poured in torrents from the smooth dome-shaped summit.

We rode on to its base and camped in a deep chasm in the western face by the side of a small water-hole. After three days' travelling without water over heavy Porcupine sandhills it was no small pleasure to watch the horses drink their fill, and it was also somewhat of a relief to find that there *was* water and that we could ourselves afford to drink without stint. To fully appreciate a wash also one

---

[*] Its total height above sea level is 2500 feet.

needs to have travelled for two or three days at least in hot weather over sandy country without any water to spare for such a purpose. Perhaps after a certain length of abstinence the desire to wash passes away, but we had just had long enough away from water to make us appreciate it from this point of view to the full.

Giles in "Australia twice traversed"* speaks of there being *permanent* water at Ayers Rock. Mr. Cowle, who had previously visited the Rock, found the water supply at the time of our visit considerably diminished, in fact no water was coming down from the rock, and it could only be a question of time as to when the two or three already small but fortunately sheltered holes around the base would be completely dried up.

PORTION OF AYERS ROCK SHOWING A DETACHED COLUMN.

In a dry season it would be very unsafe indeed to rely upon finding water at Ayers Rock, the nearest permanent pool to which is some eighty-five miles away to the north (in a straight line) in the George Gill Range, across the desert sandhill country in the midst of which lies Lake Amadeus. If the latter were not passable then a long *détour* would have to be made to get round its eastern end.

Our camping ground was in the deep chasm referred to before, and lying down in the open at night we could see just a small patch of sky overhead, shut in by the rocks which overhung so as to form almost a funnel, narrowing from below upwards.

---

* Vol. ii., p. 62.

The next day we spent quietly in the neighbourhood of the rock. Seen from the distance it looks like a great solid mass, but when close to it there are found to be a number of huge, bluff like masses which stand out each with its smooth, rounded summit melting above into the main central mass. The weathering has made the surface curiously smooth. The rock peels off in thin flakes, but at the same time weathering on a larger scale is taking place. Close to our camp was a great curved column (shown in the accompanying illustration) two hundred feet high and eight feet wide, separating off from the surface the contour of which it followed. Except at its upper and lower ends it was quite free from the rock and looked like a huge flying buttress. In course of time it would slip down and break up into big masses like those which were everywhere lying round about the base of the mountain.

In parts also the face had weathered so as to produce a curious netted or curtain appearance due to the presence of a network of more resistant material in the arkose sandstone of which the rock is formed. Small caves were plentiful in parts and the walls of these had been ornamented by the natives with drawings of hands and human faces and various animals. Some of the latter such as those of snakes and dingoes were recognisable, whilst others were apparently only conventional patterns, such as intertwined or continuous curves not without artistic feeling and suggesting the rudiments of designs which might in course of time become developed into elaborate, interlaced ornamentations. The colours were the usual red, yellow, black and white, and nowhere did we see any trace of blue such as has been described as occurring in native drawings from further north.

After spending the morning in wandering round the rock, photographing it and copying many of the drawings, a number of which are reproduced in the Anthropology report, we went out into the Mulga scrub in search of honey ants. Evidently this is a favourite hunting ground of the blacks, as the scrub was in parts thick with mounds of earth which they had thrown up when digging out the nests. A native woman armed only with a yam stick will dig down to a depth of a few feet in a surprisingly short space of time, breaking up the earth with the stick held in the right hand while in the left a small pitchi is held and used as a shovel to clear the loosened earth away.

The honey ant nest is not indicated on the surface by any mound. There is simply a hole perhaps an inch or more in length, and from this the central burrow which is about three-quarters of an inch in diameter runs down vertically with horizontal passages leading off at intervals after a depth of perhaps two feet has been reached. In the nest which we dug up during that afternoon, a few honey ants were found in each of these horizontal passages. They are quite incapable of movement, their small bodies looking like little appendages of the swollen abdomen, which has the appearance of an almost transparent bladder with the hard terga and sterna forming dark bands across it on the upper and under surface.

When the nest was disturbed the workers made no attempt to hide the honey ants, in fact it would be a matter of considerable difficulty to move these as the burrow is not large enough to allow many ants to work at once, and it would take the combined efforts of a fair number to carry off one of their honey pots.

The larger number of honey ants is apparently to be found near the bottom of the burrow, which may go down for a depth of five or six feet. Unfortunately we could not find the winged forms, but these have since been sent to me by Mr. E. C. Cowle, who has spent a considerable time in securing them under difficulties which can only be appreciated by those who have attempted to collect in such a district as Central Australia during summer time. The commonest form, the nest of which we examined, was first described by Sir John Lubbock under the name of *Camponotus inflatus* and is called by the natives Yarrumpa. The blacks are very fond of it and the women or lubras dig it up in scores from the hard sandy ground in Mulga scrub, though it is only found in certain localities.

Mr. Cowle's efforts have resulted in securing two new species during the past year, which have been described by Mr. Froggatt under the names of *Camponotus cowlei* and *C. midas*. More than thirty species of the genus, which is world wide in its distribution, are known in Australia, but there are as yet only three of them in which the curiously modified individuals are known to exist. Of *C. cowlei* I found a few specimens in a small nest under a block of quartzite in the McDonnell Range, but the splendid series since secured by Mr. Cowle at Illamurta in the James Range, includes all the various forms. The body is a golden colour with the terga and sterna orange-tinted and they do not appear to reach the size of the honey ants of the species *C. inflatus*. In *C. midas*, the specimens of which I owe entirely to Mr. Cowle, the individuals do not appear to become anything like so much inflated as in both of the other species, and the honey ant, though swollen out, is probably capable of a certain amount of movement.

Honey ants similarly modified to those found in Australia have been described by Mr. W. Wesmael as occurring in Mexico and by Mr. H. C. M'Cook in Colorado. In each case they exist in dry, arid country and the modified individuals may perhaps be regarded as specially connected with the nature of the surroundings. Just as bees store up honey in combs and use it when food is scarce, so these ants store up honey, not in combs, but in the bodies of certain members of the community. The head and thorax of the animal remain unchanged, but the crop lying in the abdomen becomes enormously inflated, and it is in this that the honey is stored up. When required for use the other ants are said to come and tap the sides of the swollen abdomen with their feet, and in response to this stimulus the honey is passed in drops out of the mouth of the modified honey ant and is then eaten by the others.

In the water-hole close to our camp large tadpoles were swimming about. There was no sign of any adult frog to be seen or heard, but of course they must have been in hiding somewhere not far off as there was no permanent water within more than eighty miles and no frog could possibly live in the sterile desert country which stretches far out in all directions around the Rock, and the distance was far too great for the eggs to have been carried by birds. The tadpoles belong to the species *Helioporus pictus*, which in these parts, as elsewhere, burrows and so can afford to wait quietly during the intervals, often lasting several months, which elapse between successive rainfalls and during a part of which time—how long will depend entirely upon the length of the drought—there can be no water at Ayers Rock.

The water sheltered in the deep chasm was so cold that the rate of development of the frogs would be very slow, and this very coldness of the spot and consequent slow rate of evaporation would allow a much longer time for development than if the water were more exposed and consequently evaporated more rapidly, as it does in the pools on the open Steppe lands.

Towards evening we climbed a little way up the face of the rock at the solitary spot where the slope is sufficiently gradual to allow of an ascent being made for even a short distance. The only white man who has ever scaled the rock is Mr. Gosse, and the climb is at best a perilous one as the least slip is fatal, and there are only thin scales peeling off which afford any surface rough enough for holding on to by either hands or feet. How steep the slope is can easily be realised by reference to the illustration.

We were looking out to the west : at our feet the sandy plain was dotted with thin scrub and away in the distance it was crossed by dark patches, where mile after mile, the thick Mulga scrub stretched across. The level line of the horizon was only broken by the great, dome-shaped masses of Mount Olga, behind which the sun was setting. For a short time the harsh features of the desolate plains were softened by the warm colours of the after glow, and Mount

Olga stood out as a purple mass in strong relief against the orange sky. It was a scene perfectly typical of the Australian desert at sunset, and to complete it as we looked down we saw a family of the native sandhill blacks making their way round the base of the mountain towards our camp.

On reaching the latter we found that our black boy had come across the family, which consisted of a man, two women and several younger ones, out in the scrub and had brought them in. None of them had ever seen a white man before and the women were in a state of great fright when they saw us, but the man soon became accustomed to us and when the first shyness had worn off proved to be the most loquacious individual I have ever met. Naturally he could not realize that his remarks were perfectly unintelligible to us, but by aid of our black boy as interpreter we managed after a time to understand what he told us. Our provisions were on too limited a scale to allow of anything like extravagance, but a little fat and sugar went a long way towards establishing what, had circumstances permitted of it, would have been on his part a life-long friendship. They made their camp a little way from ours and we spent the evening after our notes were written up in questioning our newly found friend, whose name was Langkarti tukukană.

Mount Olga.

Early next morning we started off on horseback to visit Mount Olga, our black friend accompanying us on foot. The country was dry and desolate in the extreme, with alternating heavy sandy ground covered with Porcupine grass and dense Mulga scrub. As we neared the mountain it was seen to consist, as shown in the illustration (Plate 8), of a large number of huge rounded masses arising from an elevated base and separated from one another by deep ravines.

We steered our course for the southern end, where there was apparently the highest dome-shaped mass, and rounding this just at sunset we turned into a magnificent ravine the sides of which rose precipitously for a height of 1500 feet.* The rocks were quite bare and of the usual red colour with great streaks of black looking just as if enormous cauldrons of molten tar had been emptied on to their rounded summits and had flowed down the sides. The black and red were relieved, here and there, by large patches of green apparently due to lichens growing on the surface.

---

* More than 3000 feet in all above sea level.

At the entrance to the ravine were two trees marked respectively $\frac{T}{5\text{-}75}$ and $\frac{T}{7\text{-}89}$ and this showed us that we were in Tietken's old camping ground. We came across the remains of boxes which were evidently the relics of the four left behind by Mr. Tietkens in 1889; there was no trace to be found of the camel pack-saddles which he left, though possibly a bar of iron which we saw in a black's camp near the rock may have been obtained from them. In the scrub we saw on the hard sandy ground undoubted traces of his camel tracks.

A gum creek flows away into the sandhill country out to the south-west, but here again there is no such thing as permanent water; there was evidently much less than when Giles and Tietkens visited the rock. The blacks assured us that the only water anywhere at the time of our visit about Mount Olga was to be found where we were camped, and leading our horses over the smooth rocks forming the floor of the rapidly narrowing ravine, which was in all about a quarter of a mile in length, we came upon a solitary small pool. It was just a rock-pool with no permanent supply and evidently could not last much longer unless replenished by rains, and we were thankful to find enough to water the horses. It is of course quite possible that there were other small pools which the blacks discreetly said nothing about, but there is certainly no permanent water; the pools described by Giles and Tietkens were at the time of our visit either dried up or much smaller than when they visited Mount Olga, at which time a stream was flowing over the rocky bed and out on to the loamy flats beyond, but now except just for this one pool everything was as dry as possible.

In certain respects Mount Olga is almost more impressive than Ayers Rock : it has the form of a number of huge masses like the latter thrown together and separated from one another by deep ravines. The rock in each case has weathered into smooth, dome-like structures, and these rise perpendicularly to a height of 1500 feet directly above the flat plains which surround them, so that they appear to be much higher than they really are.

Unlike Ayers Rock, which is composed of sandstone which has undergone a considerable amount of metamorphosis until now it has a striking resemblance to a granite, Mount Olga is made up of a coarse conglomerate and is probably younger in age than Ayers Rock.

Whilst riding across between the two we had suddenly emerged from a belt of scrub into a patch of more open ground and came upon a small camp of blacks living in their "wurlies," each of which was simply a lean-to made of branches which served as a protection from the weather. These sandhill blacks had never seen a white man before, and in their alarm one or two of the men seized their spears and poised them on their womeras or spear-throwers, but fortunately Laungkartitukukana's powerful voice was heard just in time to prevent what would have been an uncomfortable reception for ourselves. They evidently thought that man and beast were one creature, and when the latter came in two and we dismounted they were much alarmed and sat down huddled together—the women and one or two of the younger men crying from fear. However we reassured them as well as we could and they promised to come to our camp, but as soon as we were out of sight in the scrub they took all their worldly possessions and fled up one of the lower hills flanking the main mass, and there as the darkness came on we saw their camp fires dotted about.

Much to our regret we had no time to explore the mountains, as our arrangements left us only just time to join the main party at Glen Helen in the McDonnell Ranges at the date previously fixed upon. After taking a photograph of the ravine, the light for which was unfortunately very bad, we started back towards Ayers Rock.

A great deal of persuasion and shouting was necessary in order to bring the blacks down from the mountains. Laungkartitukukana exerted himself to the uttermost, and the contortions of his body whilst he forced out a volume of high pitched sound were most remarkable.

At length we saw them coming down and after treating them to a little sugar and fat and the remains of very hard "johnny-cakes" and presenting them each with a few matches—a valuable present—and a little tobacco they became reassured. The men wore the emu feather "chignons," frequently seen in this part of the country. These are pads about ten inches in length, six in breadth and two in thickness, made up of emu feathers matted together in much the same way as in the Interliña or feather shoes worn by the Kurdaitcha. They are tied on to the back of the head with string made of opossum fur, and into the upper angle on each side is stuck a skewer of wood with a little tuft of the white tips of the rabbit-bandicoot (*Peragale lagotis*) tails, which they call Alpita, or else a tuft of feathers, often of the Eagle-hawk. In addition to the opossum string the chignon is attached to the hair by means of sharp-pointed pieces of wallaby and kangaroo bone.

One man was carrying a small bag of skin, probably of the rat-kangaroo, tied round with hair string and containing, apart from his girdle, shield and womera, his worldly possessions. These consisted of a tuft of emu feathers for corrobboree purposes, a few odd bits of flint and pieces of kangaroo and emu tendon and also a rather fine tuft of Peragale tail tips belonging to his wife and forming her dress and ornament on special occasions. As his wife was not with him and he had evidently considerable misgivings as to what might happen if without her consent he parted with her belongings, I had great difficulty in persuading him to barter the little bag and its contents and had eventually to part with my sheath knife to secure it. It was in this camp that we found the iron bar previously referred to, which, as the blacks had never seen a white man before, had probably been taken from Tietken's pack saddle and adapted as a yam-stick, though of course it is quite possible that it might have been traded down from further north.

After the blacks had become friendly three more of them insisted upon following us across to Ayers Rock, and without any difficulty at all kept up with us for the whole of the twenty-two miles which we traversed through the scrub and sandhills. On the way over we photographed the Range, the eastern face of which must be fully five or six miles in length, and set fire to tracts of Porcupine grass. As soon as these are ignited the hawks assemble, though none are to be seen before, and pounce down upon the smaller animals such as lizards which are driven out of the burning grass.

After reaching our old camp at Ayers Rock there was still an hour or two of daylight left and Mr. Watt and myself went out to make a further investigation of the caves with their native drawings. On the rocks we found a few fig trees and Acacias and the Gastrolobium plant growing, which in certain parts of the Central district is very destructive to cattle which readily feed upon it and are poisoned. All round the base of the western face of the rock grow very fine specimens of *Acacia salicina* with its light green drooping foliage, and the kangaroo grass reaches a height of six feet or more.

As we wandered back to camp at sunset the scene was exceptionally beautiful. On the ground the tall, light yellow grass and the green Acacias stood out in strong contrast to the venetian red rocks which rose perpendicularly for nearly a thousand feet, and above them was the cold steel-blue sky.

The blacks had again made their camp close to ours and during the afternoon they had secured two kangaroos out in the scrub. The kangaroo was the common red one (*Macropus rufus*) which has evidently a wide distribution as it is apparently the only species inhabiting the plain country, and as the same class of country stretches right away into Western Australia presumably the species is distributed throughout it.

It was by this time quite dark in the chasm and the blacks were preparing for their feast. It was a strange sight to watch these natives, who were in a genuinely wild state, none of them having seen a white man before. Sitting round their fires two of the men prepared the kangaroos for cooking. First of all the two large tendons were extracted from each hind limb. To do this the skin is cut through close to the foot with the sharp bit of flint which is stuck on to the end of the womera or spear-thrower by means of the resin obtained from the leaves of the Porcupine grass. A hitch is then taken round the tendon with a yam stick, and then with one foot against the rump of the animal they pull until the upper end of the tendon gives way. Then the loose end is held in the teeth and when tightly stretched the lower end is cut through with the flint and the tendon thus extracted is twisted up and put beneath the waist girdle for safe keeping. These tendons are of great use to them in various ways such as that of attaching the points on to the end of their spears and womeras or for binding round the splicings on their spears. After this is done a small opening is made in the abdomen wall with the flint and through this all the intestines are pulled out and cut off. The hole is stitched up with a short pointed stick, the limbs are dislocated, the tail cut off at the stump, and then the animal is ready for cooking.

The women and children took the intestines and at once cooked them by means of rubbing them continuously in the hot sand and ashes. Meanwhile some of the others had dug with yam sticks a shallow hole in the ground just large enough to hold the body and had made a fire in it. When this had burned down and nothing was left but hot ashes the kangaroo was laid on the latter, some of which were also scattered over it but not so as to cover it entirely. After lying here for an hour it was supposed to be cooked and was taken out and placed on Acacia branches. It was then cut open and first of all the liver and heart were taken out and eaten. The carver took the burnt skin off often using his teeth to tear it away and with a yam stick cut the body up roughly into joints, helping himself as he went along to such dainty morsels as the kidneys. Everyone, women and children included, had their share of the meat, and if not done enough it was well rubbed in the hot sand and cooked therein to suit the taste of the eater. There did not appear to be any special portions given to any individual, but the men were served before the women and the children received pieces from the men and women. It was by no means an appetising sight and the whole method was very crude, and nothing like so much care was taken in the cooking as is often the case amongst other Australian natives, who make a deep hole and cook their game on hot stones, the former being completely covered with earth, from actual contact with which they are protected by green leaves during the process.

These natives living amongst the sandhills and bare ranges of Central Australia are in certain respects amongst the lowest of the Australian aborigines. They make no use of the skin of the kangaroos and wallabies, which are by no means uncommon, for purposes of clothing; they have not even any netted "dillybags" such as are made by natives in other parts; their weapons and implements of various kinds are of the simplest nature and but little ornamented, and such designs as they do carve or paint upon them are very crude when compared with those of the northern tribes. Their stone implements are interesting because they are simply chipped and no attempt is made to grind them down so as to produce smooth surfaces. So far as these are concerned they are a strong contrast to the ground axe-heads made by the natives all along the coastal district from Victoria northwards.

When they had eaten as much as they could they laid themselves down for the night and all was quiet, except for a minute or two every now and then when one or other of them woke up and raked together the embers of the fires around which they slept.

The Return to the George Gill Range.

Next morning we started north to retrace our steps to the George Gill Range. The first day brought us to Coulthard's Well, or Kurtitina. Leaving this at sunrise we reached Lake Amadeus at eight o'clock, and after photographing crossed its salt bed once more.* The surface was covered in parts with numberless little cones about half-an-inch high and the same in diameter. A circle of dark sand grains about three inches in diameter surrounded each, everything else but this thin circle being quite white with salt. A small hole in the cone led down into a vertical passage from one-and-a-half to three inches in depth. Each contained from two to five small, black, winged hymenopterous insects which were alive but quite quiescent, probably because of the cold. Unfortunately the specimens which I collected got spoilt during the day's rough riding, so that they cannot be determined. These and a solitary spider walking on the surface were the only signs of animal life and the Lake was as silent and deserted as when we first crossed it in the dusk. Not even a solitary bird was to be seen.

---

* The bed of the Lake is 1380 feet above sea level; that of Lake Eyre being 30 feet below sea level.

After crossing on foot we pressed on to Kamaran's Well hoping to find the water fit for the horses to drink. Another dingo had fallen in but it had been pulled out by the blacks who had evidently visited the spot during our absence, and had tried to burn the dead bodies which we had previously pulled out. As the horses were very thirsty some of them after a considerable amount of persuasion drank a little out of a sheet of canvas, the odour of which reminded us for many days of Kamaran's Well.

That night we camped amongst the sandhills and had as on the previous one to sit up and watch the horses to prevent them from wandering away in search of water. It was so cold that our water bags were frozen solid at daybreak, when we had our breakfast and started off.

The third day brought us at evening to the George Gill Range and the welcome waterhole at Bagot Creek. We were much struck with the fact that during the two weeks which had elapsed since we were here the water had very considerably increased in volume. The only explanation of this can be that, except in very dry seasons, a constant though small supply comes down from the hills and that in comparatively cool weather the evaporation is not great, and therefore though no rain falls the water holes increase in size.

The next day was a rather hard one for the horses, as we had to take them over the George Gill Range and down into a valley through which the Petermann Creek flows. The upper part of this valley, which is known as Petermann Pound, forms a large, roughly-circular flat some three miles in diameter and completely shut in by hills except for a small outlet at the eastern end where the creek flows away.

To the south it is bounded by the George Gill Range and to the north by the Station Range, already referred to as forming the escarpment on the south side of the Tempe Downs Valley. At the western end of the Pound these two ranges curve over towards one another and unite together. After traversing the flat we ascended the Station Range and at last, after the horses had had a very rough time clambering over and amongst the rocks, we came down into what is called Shakes Plain at a point some twenty miles to the west of Tempe Downs Station, which was reached after dusk.

We found that Mr. Thornton had gone away, in fact there was only one white man left in charge and he was by good fortune the cook. Just as before we were most hospitably entertained and our recollection of Tempe Downs will be of the most pleasant kind.

After half a day's spell here we left and retraced our steps through the Walker Gorge and then on to the Illara water-hole on the Palmer River. From this point, instead of going eastwards towards the Finke at Running Waters, we turned north and followed the Palmer. The river was running in a somewhat wide gorge which every now and then widened out into a scrub and grass-covered flat amongst the hills. Along the river were a series of small water-holes on which were a solitary pair of shags and a few black duck and teal. As usual, a yard away from water all was dry and the bed was marked with patches of salt.

We camped for the night in the valley after a short day's travel of only seventeen miles. It was perhaps our coldest night—at least we felt the cold most severely—as we woke at 5 a.m. to find our water bags frozen solid and the thermometer registering 16 F., at which point it remained until sunrise. As usual we camped in the open, going to sleep by the side of a fire and as there was a slight hoar frost we felt the cold far more than in the perfectly dry parts. Dressing in the open air, when the temperature is sixteen degrees below freezing point, is accomplished as soon as possible. An hour or two after sunrise we were glad to take advantage of any cool shade.

All day long we followed the windings of the river, sometimes through open flats, sometimes along deep, rocky gorges several miles in length where high precipices rose on either side directly from the narrow river bed, and where it was difficult amongst jagged masses of rock of all shapes and sizes which blocked the bed of the stream (or rather would have done so had there been any stream) to find a safe footing for the horses. When the river is actually flowing the passage of these gorges is quite impossible, but during the dry season they contain no water or at most only a few small pools.

Emerging from the last gorge we halted to give the horses a rest by the side of a pool known as Bowson's Hole. On a high rocky bank in the last gorge we saw the largest tussock of "old man Porcupine" grass which we met with during the Expedition. Its height was at least seven feet and its diameter fifteen feet. As usual in all the older tussocks the central part had died away. By the water-hole, under stones where the sand was moist, were as usual numbers of a little brown carab beetle (*Tachys spenceri*) which proved to be a new species, but which is very common in this situation by the side of all the water-holes in the James and McDonnell Ranges. In the water-hole itself Vallisneria and Chara were growing with a few snails such as *Isidorella newcombi* and *Ancylus australicus* creeping about on them in abundance. The latter has a little shell not more than the sixteenth of an inch in length looking exactly like a minute limpet and it is widely distributed in Australia.

There was but little insect life to be found though the Cassias as usual were brilliant with yellow blossom. A black and white mason fly was making persistent efforts to drag a heavy spider up the smooth trunk of a red gum to its nest. The spider was apparently too heavy for it to fly with and was struggling, but walking backwards up the trunk the insect tried to drag the spider after it, and would doubtless have eventually succeeded had it not been transferred with its prey to the collecting bottle.

After the mid-day halt we still followed the Palmer up, but now in a valley with the ranges gradually receding on either side until shortly after sunset we reached a spot close to the source of the river where the hills on the one side ran away southwards and on the other towards the north-east, the valley itself opening out westwards into the broad Missionary Plains.

Here close to the base of a projecting, somewhat conical hill known as Pine Point we camped for the night.

Early next morning we passed over a low rise separating the Palmer valley from the plains. The latter form a long stretch of country some twenty miles in breadth which runs east and west. On the north they are bounded by the McDonnell Ranges and on the south by the James Range, which towards the east run in a north-easterly direction so as gradually to narrow in this end of the plains. Westwards they stretch away to open out into the desert sandhill country lying beyond the mountain ranges.

As soon as we came upon the plains we found ourselves in a belt of grass trees (Plate 4) belonging to a species not hitherto described. The first specimens with which we came in contact were shown to Professor Tate along the valley of the Palmer, some few miles north of the Illara Water-hole, by Mr. Thornton of Tempe Downs, after whom the species has been named (*Xanthorrœa Thorntoni*). They seem to stretch in a narrow belt some seventy miles in length right across the plains as far as Glen Edith in the neighbourhood of which they were again seen by Professor Tate. The larger specimens have a stem some five or six feet high with a crown of long wiry leaves and a flowering stalk the top of which is fully twelve feet above the ground. In general appearance they are much like the larger grass tree (*X. major*), but can easily be distinguished from this by their much more linear leaves. They form a good example of several species of plants which in this central district have a very limited distribution. Amongst these *Swainsonia canescens* was only found in two small colonies nearly sixty miles apart, and *Goodenia Horniana* in the same way was only met with in two spots a hundred miles apart. Of course these and other such plants may occur elsewhere, but as a constant watch was kept every day over the large area of country traversed it is quite safe to adopt Professor Tate's opinion that they are extremely sporadic in occurrence. Just like certain of the animals such as the earthworm and various species of snails those which still persist may be regarded as relics of a former more widely-spread flora which have, under the gradually increasing desiccation of the country, been able to persist in favourable spots. After photographing the grass trees, a group of which with Pine Point in the background is represented in the illustration (Plate 4), we rode on over undulating country with a broad belt of scrub rather richer than any we had hitherto seen as it contained plenty of Prostanthera, various species of Eremophila in flower and Currajong trees. The Mallee Gum was thickly covered with a bright red flowering mistletoe. To the north-east we could see the Gosse Range, an isolated mass about two miles in length and the same in breadth. The country all round this was thick with Porcupine grass amongst which were fine specimens of *Acacia dictyophleba* with large yellow balls of flower, and the ground was cut through by deep, narrow watercourses down which in rainy seasons the water pours from the hillside, only to become rapidly lost.

Turning round the western end of Gosse Range we struck Rudall Creek, which runs east from here to join the main Finke. Our camp at night had been a dry one, so that we were glad to find a small water-hole. It lay in the creek bed at the base of a rock; probably beneath the sand a bar of rock runs across and so causes the water to come to the surface. While we rested a flock of rock pigeons (*Lophophaps leucogaster*) came down to the water-hole. These are amongst the most distinctive birds of the district; in colour they resemble, generally speaking, the yellow-brown sand or rock on which they remain quiet until you are close to them when they rise with a whirr and then, once on the wing, glide away quietly. They have a curious habit of making a kind of run down to a water-hole. In this spot, for example, the pool was hemmed in on one side by a rock about fifteen feet high, while on the other side was a level sandy bank. The birds congregated at the top of the rock and then one after the other ran down a beaten track to the water and up again.

From Rudall Creek we travelled north towards what looked like a series of rounded, smooth, grass covered hills much like the Downs of the south of England.

When we were amongst them we found however that they were a series of jumbled hills covered all over with Porcupine grass, the tussocks being so close together as to give when seen from a distance the appearance of a smooth carpet of grass, but which in reality made travelling somewhat slow and uncomfortable.

Beyond this low range we could see the peaks of higher hills, but it was dusk before we had made our way up and into the near Range where we camped for the night.

In the morning we passed through a narrow cleft in the hills and struck the deserted Glen Helen Station at the base of Mount Zeil within half an hour of the time at which the main camel train had reached it. It was just a fortnight since the two parties had separated at Reedy Creek and during that time we had traversed some three hundred and thirty miles. We had actually been travelling for twelve days, as a day and a half had been spent in camp at Ayers Rock and half a day at Tempe Downs, and as there had been two spells of three days each over waterless country and our journey had lain almost entirely over heavy Porcupine sandhills and across rough, rocky ranges our horses had had by no means an easy time.

We found that the main party with the camel train had travelled from our parting place at Reedy Creek eastwards to Carmichael Crag which forms the eastern end of the George Gill Range and had then turned northwards across the open country to Glen Edith, and then travelling westwards had struck the eastern end of the McDonnell Range near to Haast's Bluff. After traversing for some little distance the western end of the narrow Horn Valley which stretches in an unbroken line for some two hundred miles eastwards and which, as at the Mereenie Bluff, is hemmed in by very fine escarpments of rock often rising for several hundred feet vertically, they crossed the valley and travelled northwards towards the Darwent Creek. Turning south again they then passed along the valley lying to the south of the main McDonnell Range and so reached the base of Mount Sonder.

It was during this part of the journey that the only specimens seen of the rare Princess Alexandra Parrakeet (*Spathopterus (Polytelis) alexandrae*) were secured by Mr. Keartland. Near to Glen Edith a flock of these birds was found in a patch of Desert Oaks. Their long slender tail and delicate tints of green, blue, purple and salmon-pink render them perhaps the most beautiful of our Australian Parrakeets and up to the time of the Expedition, though they were first discovered by Waterhouse on Stuart's Expedition into Central Australia, only a few specimens had been secured.

Mr. Keartland was fortunate enough to obtain fifteen and since that time though they were very rare indeed before they seem from some cause to have appeared in considerable numbers at one or two spots, such as the Hale River to the east of Alice Springs and at Illamurta in the James Range during the early summer months (November) of 1894, but since then they have again disappeared.

This sporadic appearance both in space and time of various forms of animals is very characteristic of many Central Australian species. Perhaps for the space of a month, owing doubtless to the occurrence of a combination of favourable circumstances, an animal will suddenly become abundant and then as suddenly again become rare, only to reappear after the lapse, it may be, of several seasons.

The presence of a peculiar, spatulate, third primary feather in the wing of the adult male has caused Mr. North, in whose hands the birds secured during the Expedition were placed for description, to separate the species from the genus Polytelis in which it was placed by Gould and to place it in a new genus to which, in allusion to the presence of this peculiar feather, he has given the name of Spathopterus.

Its food evidently consists mainly of grass seeds and according to Mr. Pritchard — one of the prospectors accompanying our party — who has seen a considerable number of specimens since our return the birds nest in hollow trees, often several pairs occupying one tree, and lay five eggs in a clutch. Mr. Keartland experienced considerable difficulty in distinguishing the birds owing to their curious habit of "lying along the stout limbs of the tree like a lizard," instead of adopting the style of most other Parrots and perching on a twig or thin branch. Mr. Pritchard however writing in November, 1894, to Mr. Keartland said: "This is the first time on record that they have made this (*i.e.*, the Hale River to the east of Alice Springs) their breeding ground, but I do not think that they have come to stay, and perhaps in a year or so they may be as rare as ever. . . . They live in hollow trees, laying five eggs in a clutch, and several pairs of birds occupy holes in the same tree. They are nesting now in the Eucalypts on the banks of the Hale River and other large watercourses. They do not always lie along the limbs as you found them at Glen Edith, but perch as other Parrots. I have a number of them in captivity, amongst them being an old male bird with a tail seventeen inches long."

CHAPTER VI.

The Higher Steppes.—The McDonnell Ranges.

Camp at the Base of Mount Sonder—The Redbank Creek and Gorge—Description of Fish found in the Waterholes—The Horn Valley—Origin of the Gorges—Camp in the Finke Gorge—The Hare-Wallaby and Rabbit Bandicoots—Travel South along the Finke and across the Missionary Plains to Hermannsburg—The Mission Station and its Influence on the Natives—Divide into Three Parties—Follow the Finke through the James Range to Palm Creek—Three Days Camp at Palm Creek—Palms and Cycads—Account of the Animal Life of Palm Creek—Restriction of Species to a Small Area as exemplified by the Mollusca—Return to Hermannsburg—Jerboa Rats and Antechinomys—Leave Hermannsburg—Modification in Form and Colour of the Foliage of Acacia salicina and Mulga—Camp in the Scrub—The Main Camel Team goes on Eastwards along the Missionary Plain to Alice Springs—A Section of the Party goes North to cross the Ranges to the Burt Plain—View from the South McDonnell Range—Camp near Paisley Bluff—A Day in Camp—Various Forms of Ant Nests—Rock Wallabies—Method of Carrying the Young in the Pouch, a Severe Handicap to Marsupials in Competition with Rodents—Brinkley Bluff—Traverse the Ranges and Camp on the Burt Plain—Strike the Telegraph Line and follow it South to Alice Springs—Mr. Watt pays a Flying Visit to the Gold and so-called Ruby Fields—A New Marsupial—The Ranges at Alice Springs—The Todd River—Coolin Lagoon—Various Forms of Phyllopods and their Habits—The so-called Barking Spider—The Sound probably due to a Bird—The Presence of a Stridulating Organ in the Spider—Leave Alice Springs and travel South along the Telegraph Line to Oodnadatta.

FROM Glen Helen Station, which was quite deserted and in ruins, we went a few miles further east and camped close to the base of Mount Sonder. We were at length in the real McDonnell Ranges, but they were very different from what on starting we had expected to find. Bare peaks, some of them nearly 5000 feet high, rose at intervals abruptly from amongst a mass of low ridges flanked, especially to the north, by jumbled hills. Here and there creeks forced their way across them through gorges cut deeply in the rocky ridges, but there were no great sheltered valleys or luxuriant vegetation; everything was bare and dry except for the gums bordering the creek beds and the porcupine grass, patches of which extended even to the tops of the highest peaks.

These peaks are situated in what Messrs. Tate and Watt recognise as the Pre-Cambrian area. In various parts, such as the Belt Range and Mount Zeil, they consist of quartzite capping an underlying mass of Pre-Cambrian gneissic rocks which form the jumbled hills stretching north towards the Burt Plains. Mount Sonder, near to which we were camped, was formed of Ordovician quartzite, but in the valley of the Davenport Creek, close by its south-western base, gneissic granite was seen outcropping and representing in all probability an inlier of Pre-Cambrian rocks. Its southern base was flanked by low limestone hills and about a mile to the south of us across the small alluvial plain along which the Davenport and Ormiston Creeks, forming the main sources of the Finke, were running, rose a long ridge of Ordovician Quartzite.

Climbing over the limestone hills a little distance to the north of our camp we came upon the Redbank Creek, running south, just as it emerged from the gorge in which it passes through the lofty quartzite ridge from which, immediately to the west of the gorge, rises Mount Sonder.

The watershed lies well to the north of the line along which are now the highest peaks such as Mounts Sonder, Zeil, Heuglin and Giles, and the creeks have in course of time cut their way in deep gorges through the ridge which forms the southern boundary of the Pre-Cambrian area.

The accompanying diagram will serve to give a general idea of the main physiographic features of the region. It is supposed to represent a section cut from north to south from the Burt Plains, which lie to the north of the main McDonnell Range to the James Range in the south. Starting in the south we find the Missionary Plains, which vary considerably in width, gradually narrowing from about twenty miles at the western end to perhaps a mile or two at the eastern end in the neighbourhood of Alice Springs.* Going north across these we come to a series of low hills and then cross a distinct ridge with a steep northern escarpment and so descend into the Horn Valley, which is at most only about a quarter of a mile in width. Crossing another distinct ridge bounding the Horn Valley on the north we come into another broad valley, perhaps half a mile across, lying at the base of the main McDonnell Range. The latter consists of a series of low jumbly hills with a main ridge in the southern part the whole running east and west for some 100 miles. This general arrangement of parallel valleys and ridges all running east and west, with the main river channels cutting across them from the watershed in the north, is the striking physiographic feature of the Higher Steppe region.

---

* They are not called the Missionary Plains except in the broad part out to the west end, but the valley is really directly continuous from east to west and is gradually narrower in eastwards as the James Range (here usually called the Waterhouse) trends north-east so as to approach the McDonnells.

As we followed up the Redbank towards the mountains the bed narrowed and the rocks closed in on either side until we came to a deep pool lying at the entrance to a gorge, which was not more than six feet wide.

For half a mile this gorge which is nothing more than a zig-zag cleft cuts its way right through the range. Its narrow bed is filled with water, deep and intensely cold and on either side the red jagged rocks of quartzite rise precipitously for several hundred feet. In contrast to the open valley and plain across which the river flows as soon as it has forced its way through the mountain, the deep cleft with its still waters and its rocky sides forms a most impressive sight.

Some idea of the nature of this gorge may be gained from the illustration which is reproduced from a photograph taken at mid-day during the short interval of time when it is lighted up. It is of course impossible in a photograph to give any adequate representation of a scene which depends for its effect upon rocks brilliant red in colour, a deep rock-pool and a cleft through which can be seen a narrow strip of bright blue sky.

Such gorges of which this is perhaps the narrowest and most confined form one of the most striking features of the McDonnell Range across which they always run from north to south. In all probability they owe their origin to the fact that the streams which now flow through them were able to keep pace with the gradual elevation of the mountain ridge, the streams wearing out the gorges as the land rose. In some cases as in those of the Palmer River already referred to, and still more strikingly shown in the case of the great winding gorge through which the Finke flows in its passage across the James Range they may be many miles in length.

They afford the only means of traversing the ranges which run continuously from east to west as the rocks are far too steep and jagged for the passage of horses and camels. Sometimes after a heavy rainfall the water will scour out the bed of the gorge and transform it from a dry track into an impassable water-hole.

It is upon the shady sides of these gorges that many of the most characteristic Larapintine plants, that is those of the Higher Steppes have found shelter, and it is in them also that the water-holes are really permanent and here also must live the fish which in times of flood are carried away to the south to stock the water-holes along the rivers which rise in the McDonnell Range and flow south across the Lower Steppes.

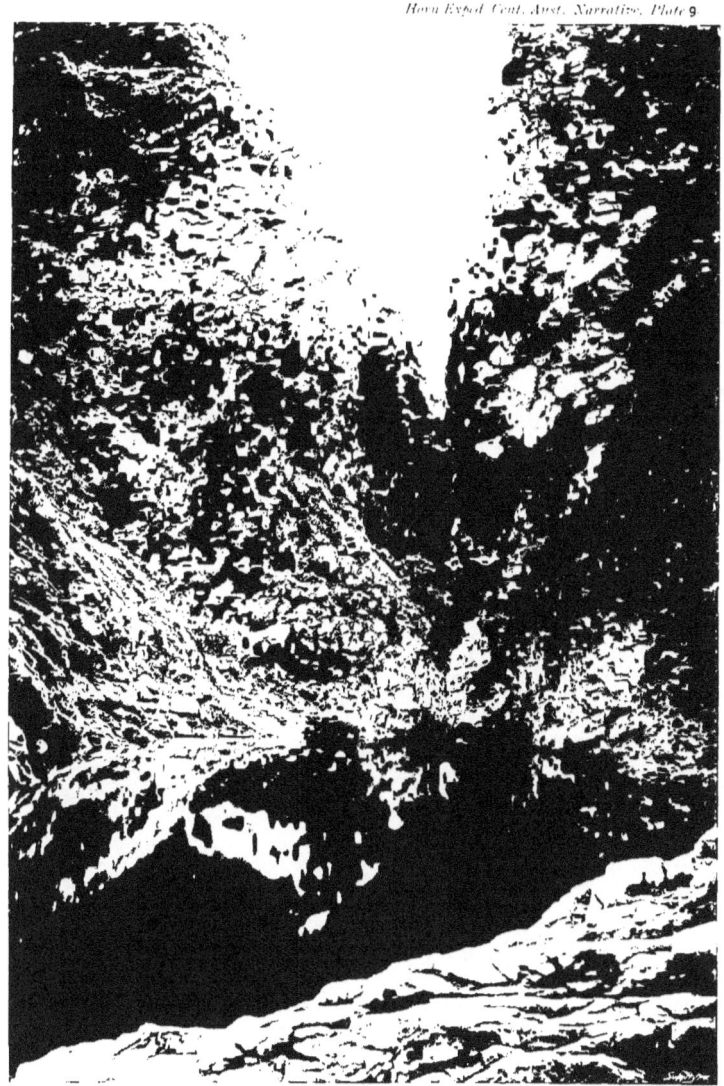

RED BANK GORGE.

The few rocky pools lying in the more open part of the gorge, though none of them were more than a very few yards square were well-stocked with fish, and out of the eight species met with during the Expedition six were caught in one small pool at the entrance to the Redbank Gorge. These were (1) the bony bream (*Chatoessus horni*) the largest fish of the district, though the specimens here were small when compared with those caught in the big water-hole at Henbury, (2) a large species of Therapon (*T. truttaceus*) silver-grey in colour with golden spots, the specimens of which were the largest of this species caught during the Expedition, (3) a smaller species of the same genus (*T. percoides*) easily distinguished from the former by its bright silver colour and by the presence of five strongly marked dark bands running vertically on each side of the body, (4) and (5) two small, thinner fish, closely allied to one another (*Nematocentris tatei* and *N. winneckei*) with golden lines running horizontally along the side, and (6) a small but more stoutly built fish (*Eleotris larapinta*) with the body a general yellow-brown colour with some ten darker vertical bands on each side.

The first five species were swimming about together, and here as elsewhere when the fish went together in a common shoal the most prominent was the little *Therapon percoides* with its silvery body and black bands, but it was also the quickest in its movements and the most difficult to catch. The water in these rock-pools was always perfectly clear and the only way to secure the fish was to drive them into a narrow part of the pool if there happened to be one and then to use the hand net. The little Therapon when taken out the water made a small but distinct trumpeting noise. The Eleotris did not often swim about with the others but lay near to the bottom of the pool, usually in fact resting on the bed where it was sandy.

In addition to the eight species collected during the Expedition and identified by Mr. Zietz a single specimen of *Therapon fasciatus* has been recorded by Mr. Lucas[*] which was secured "near the McDonnell Ranges." Mr. Lucas has also mentioned the occurrence of a species of Chatoessus which he says "seems to correspond better with *C. erebi*, Gunth., than with *C. richardsoni*, Castl.," but as this was an immature specimen he was not able to identify it with certainty and the wide distribution and large numbers of the single species of Chatoessus met with during the Expedition (*C. horni*) render it possible that Mr. Lucas' specimen was an immature one of the same species.

---

[*] Journ. Linn. Soc. N.S.W., 1891, Pt. 2, p. 362

Eight species will doubtless appear a very small number, in fact if we exclude the little *Gobius eremius* which was only found in two artificial pools in the Lower Steppes the number which is characteristic, so far as yet known, of the water-holes amongst the ranges of the Higher Steppes, whence those on the Lower Steppes are stocked is only seven. Of those secured by us all except one (*Plotosus argenteus*) are widely distributed throughout the water-holes amongst the Ranges and exist in comparatively large numbers—that is in proportion to the size of the water-holes to which except just during and after the rain season they are restricted.

Amongst the plants a certain number of additions to the collection were made the most important being *Styphelia Mitchellii*, a single colony of which was found growing high up on Mount Souder. This was of interest as it was the only Epacrid found during the whole Expedition, though of the genus Styphelia more than one hundred and seventy species are recorded from Australia, the head-quarters of the genus being West Australia which has about one hundred and ten species. This particular species is also found in Queensland.

The district was too dry to yield anything like a good harvest to either Botanist or Zoologist and accordingly after two days' spell, during which we worked hard with very disappointing results, we determined to go on to the Finke Gorge. Mr. Cowle, who had left us on his return to Illamurta, had reported that the passage of the Gorge was blocked on the north by a water-hole stretching across it, so we had to go some twenty-five miles round to get through the ranges.

The valley in which we were camped lay to the south of the irregular mass of ranges some twenty to twenty five miles in width and about four hundred miles in length, which are of Pre-Cambrian age and form the McDonnells proper. To the south of us lay what are usually spoken of as the Southern McDonnells but which are of Ordovician (Silurian) age and have in the report on the Geological work of the Expedition been spoken of as the northernmost part of the James Range, which extends southwards with a mean width of sixty to seventy miles. To the south of the James Range again and in line with one another are the George Gill and Levi Ranges.

For the sake of convenience I still use the name Southern McDonnells as applying to the two very distinct quartzite ridges which run along parallel to one another for a distance which is probably not far short of two hundred and fifty miles. Between the two ridges there runs for the whole length a valley varying in width from at most a mile to a quarter of a mile to which the name of HORN VALLEY is now given.

These two ridges and the Horn Valley between them form one of the most striking physiographic features of this part of the country. At four special places* the ridges were broken through by river gorges, and the fact that in each case the north and south ridges are both cut through in this way at points opposite to one another goes far towards demonstrating the truth of the theory that the gorges have been gradually cut by the watercourses while the ridges were in course of being elevated.

The only other way in which these gorges could have been formed was by the union of ravines which by chance lay exactly opposite to one another on the two sides of the ranges.

If the gorges were formed in this way it is a very remarkable coincidence that in the case of both the main stream of the Finke and its large tributaries, the Ellery and the Hugh, such gorges should have been formed *exactly opposite* to one another in the high quartzite ridges which now form the two ranges, enclosing between them the Horn Valley.

In no instance are any other gorges formed opposite to each other in this way, and then in addition to this the long winding gorge of the Finke—more than forty miles in length—through the James Range, which has been hollowed out by the same stream as the one which runs through the northern gorges, can only be satisfactorily explained by supposing it to have been cut by the river as the land rose.†

If we suppose the river courses to have been determined since the date of upheaval of the ridges, then it is an inexplicable feature that only minor streams should follow the trend of the longitudinal folds whilst the four main ones—the Finke proper with its large tributaries the Ellery and the Hugh and again further eastward the Todd—should run in a direction practically at right angles to the length of the ridges and that only small tributaries should flow into them from the valleys—two or three hundred miles in length—which they cut across in their course southwards to the great Cretaceous plain which gradually sinks towards Lake Eyre.

---

* There are other gorges formed, but I am here only referring to the four spots at which gorges are formed through the two ridges opposite to one another.

† The James Range is unfortunately very inadequately represented in the map. The Krichauff and Waterhouse Ranges are directly continuous with one another and form only a northern portion of the James Range. The long Finke Gorge, here referred to, cuts across the latter from Hermannsburg in the north to a little distance to the north of the point at which it is represented in the map as joined by the Ilpilla Creek.

From our camp at the base of Mount Sonder we followed the Davenport back for a short distance into the Horn Valley and then turned eastwards along the latter. In this part the valley was about a quarter of a mile wide; the ridge to the north had a somewhat steeply sloping side corresponding to the dip of the ridge, but that to the south had a high escarpment, the rocks forming which have split into blocks which have tumbled over on to one another in such a way that the appearance of horizontal stratification is produced.

After travelling some twenty-five miles we came to the main stream of the Finke running straight from north to south across the narrow valley, which was here not more than three or four hundred yards wide.

The streams which unite to form the river rise partly in the country lying to the north of the higher Pre-Cambrian hills, such as Mount Sonder, partly in the valley between the latter and the northern range bounding the Horn Valley, and partly but only to a small extent in the Horn Valley itself. The various small tributaries some of which, such as the Redbank and the Davenport, pass by deep gorges through the mountain ridges, unite together not far from the base of Mount Sonder on its southern side. After running a few miles eastward and being joined by other small creeks the main stream turns southwards and, as detailed before, cuts across the ranges one after the other.

The northern gorge as represented in the illustration (Plate 10) is a short one, and only about twenty or thirty yards in width, the entrance to it being at the time of our visit completely closed by a deep water-hole stretching right across between the rocks.*

The southern gorge was considerably larger and much wider and the sandy bed of the river was bordered by steep banks covered with scrub, behind which rose the steep cliffs. Along the sandy bed were fair sized pools of water, some decidedly brackish, others more fresh and lined with rushes. The only really fresh water was in a small spring on the steep western bank and this only held about a bucket-full of water at a time. We camped by the river on a wide, open flat in the Horn Valley, whence we could work easily in both directions as we were hopeful that the presence of the water-holes might be associated with the preservation of forms typical of the McDonnell district and not yet secured.

Though a few interesting forms were found yet on the whole the result was very disappointing. At the northern entrance to the gorge, upon the rocks, was

---

* In the map the name of this Gorge is printed "Pike" instead of "Finke."

Bory Eyed Goat Arks Nov.   Pag 10.

growing a species of Swainsonia (*S. canescens*) which was only met with in one other spot along the Todd River. Under the rocks and stones and *débris* on the hill sides were colonies of molluscs, one of which (*Angasella arcigerens*) a little snail with a series of small plate like ribs running across the whorls was new and was only found in this one spot.

The water-holes apart from the usual species of fish yielded nothing. The entire absence of frogs was very noticeable, as the rushes which grew in profusion round the water-pools might have been expected to harbour a certain number, but not one was to be seen or heard, their absence being probably associated with the slight brackishness of the water.

Around the spring on the western bank was a patch of black earth in which were found a considerable number of the earthworms previously alluded to — their cocoons, each with a well marked "tag" at either end were fairly numerous, but there was no trace of any more than the one species and they are limited to a small patch of ground only a few yards square.

Even beetles were difficult to find and amongst larger animals all that we secured were a few of the ubiquitous lizard, *A. reticulatus*, a snake, *Aspidites melanocephalus*, and a few rodents. Mammals both here and elsewhere were very difficult indeed to obtain, which was probably owing to the fact that the majority of them are nocturnal and that during the winter months, when at nights the temperature is often below the freezing point, they do not venture out. At Mount Sonder we had obtained specimens of the hare-wallaby (*Lagorchestes conspicillatus var. leichardtii*), a different species from the one found during the Elder Expedition when Mr. Streich stated that a form identified as *L. hirsutus* by Messrs. Stirling and Zietz appeared to be plentiful in the Victoria desert. We also secured in a trap a specimen of the so-called rabbit bandicoot, *Peragale lagotis*, the long, soft, grey and white fur of which renders it one of the prettiest of the smaller marsupials. Its burrows abound, being often very extensive, and it must fall a prey to the blacks in great numbers as the white tips of its tail—called by them alpita—are very extensively used to make ornaments of various kinds. They are strung together so as to form tassels, each of which may contain from twenty to thirty tails. They are to the blacks what Ermine tips have been to the whites of other parts of the world, though as fashions do not change much in Central Australia the Peragale has been more consistently sought after than even the Ermines.

The genus as at present known is represented by the species *P. lagotis* which is widely distributed in West, South and Central Australia and by *P. leucura*, only a single immature specimen of which has as yet been described by Mr. Oldfield Thomas. The exact locality of the latter species is doubtful, though probably it came from Central Australia, and Mr. Byrne who has carefully enquired into the matter thinks it possible that the blacks are acquainted with it in the neighbourhood of Charlotte Waters; but on this point he cannot feel quite certain.

Mr. Byrne has however made an interesting discovery in the form of a new species of Peragale. The specimens in question are of much smaller size than *P. lagotis* and are also of a darker colour, whilst Mr. Thomas' specimen of *P. leucura* is characterized by its almost white colour.

The new species is of about the size of a small rabbit with long dark grey silky hair; it has the characteristic long ears and white tip to the tail but the latter is nothing like so strikingly marked as in the case of the larger species *P. lagotis*. The natives distinguish clearly between the two, calling the larger one Urgätta and the smaller Urpila.

Thanks also to Mr. Byrne and Mr. Gillen a new species of the allied genus Perameles has been discovered both on the Burt Plains, near the McDonnell Ranges and at Charlotte Waters. It is evidently more closely allied to the striped bandicoot (*P. bougainvillei*) than to any other and is called "Mulgar-nquirra" by the natives at Alice Springs and "Iwurra" by those at Charlotte Waters.

The specimens of Urpila and Iwurra came from about forty miles northeast of Charlotte Waters, and Mr. Byrne has sent me the following notes with regard to their habits, and those of the Urgätta and Chœropus:—"Whilst the Urgätta occupies the inner extremity of his burrow, the Urpila during the cold weather lies within a foot or so of the entrance of his, and only uses the inner chamber during the summer. This peculiarity is taken advantage of by the natives who spring on the surface of the ground behind the Urpila breaking it in, and so cutting off his retreat to the inner chamber. He is thus compelled to rush out through the entrance where a native is waiting to give him his quietus. The Urgätta cannot be captured in this way, and has to be dug right out. Both species are nocturnal. The Iwurra and Tubaija (Chœropus) are identical in their habits, and build similar nests of grass and twigs in shallow, oval hollows scooped in the ground. They are captured in the same way, viz., by placing one foot on the nest pinning the animal down, and then pulling it out with the hand."

From the camp by the Finke in the Horn Valley we travelled south, following the course of the river across the wide Missionary Plains which lie between the South McDonnell Ranges on the north, and the Krichauff Range in the south, the latter being really a part of the James Range. Some few miles south of the gorge, Ruddall Creek, which we had previously crossed near the Gosse Range, joins the Finke.

A good track leading through the scrub showed that we were getting near to the Old Missionary Station of Hermannsburg, which we reached late in the afternoon, and where Mr. Heidenreich who was then in charge made us welcome. The mission at the time of our visit was abandoned, and the whole place more or less in ruins. A few blacks, the remnants of a larger number who were camped about the place when it was opened as a mission station, still remained, living in a squalid state in dirty whurlies. If, which is open to question, the mission had ever done any permanent good, there were no evidences of it to be seen either amongst these blacks or others whom we met with and who had been in contact with them.

The morality of the black is not that of the white man, but his life so long as he remains uncontaminated by contact with the latter, is governed by rules of conduct which have been recognised amongst his tribe from what they speak of as the "alchëringa," which Mr. Gillen has aptly called the "Dream times." Such rules of conduct are taught by the older men to the young ones and are handed down from generation to generation. Any breach of these rules renders the offender liable to severe punishment—either corporal or what is perhaps quite as bad the feeling that he has earned the opprobrium of, and is ridiculed by his fellows.

To the rules of the community the blacks, in their natural state, conform quite as strictly, in fact perhaps more so than the average white man does to the code of morality which he is taught.

To attempt as has been tried at Hermannsburg and elsewhere to teach them ideas absolutely foreign to their minds and which they are utterly incapable of grasping simply results in destroying their faith in the precepts which they have been taught by their elders and in giving them in return nothing which they can understand. In contact with the white man the aborigine is doomed to disappear: it is far better that as much as possible he should be left in his native state and that no attempt should be made either to cause him to lose faith in the strict tribal rules, or to teach him abstract ideas which are utterly beyond the comprehension of an Australian aborigine.

I do not in any way intend in saying what has gone before to suggest that the Missionaries in charge of the Station did not do their work zealously, but simply that the task which they essayed was one which under the nature of the circumstances could not be successfully carried out.

At one time as date palms and relics of plots of vegetables showed there must have been a very good garden indeed, in fact where water is available there is no difficulty in rearing vegetables, as we found by pleasant experience at Charlotte Waters, Crown Point, Hermannsburg and Alice Springs.

A little to the south of the Station across the broad valley in which the Finke was running stretched the Krichauff Range. The highest point was Mount Hermann and just to the east of this the river left the plains and entered a deep gorge, which runs for some forty miles south through the main James Range until close to Running Waters it emerges and then runs on southwards to traverse the great Desert Sandstone Plain.

This gorge was first traversed in 1872 by Giles, when he passed through it from south to north and then followed up Rudall Creek past Gosse Range and away to the west beyond Carmichael Creek.

At Hermannsburg we divided into three parties, the main camp stayed at the Mission Station, Mr. Watt with Messrs. Pritchard and Russell went out north-west so as to strike the Ellery Creek, their intention being to follow this up through the McDonnells and then travel eastwards along the Burt Plain to Alice Springs; the rest of us, that is Messrs. Tate, Stirling, Winnecke, Belt and myself, started off to follow down the Finke Gorge to the Glen of Palms.

Crossing over the plain to the Krichauff range we entered the Gorge and followed its windings for eight or nine miles between lofty cliffs of red sandstone, which sometimes hemmed in the river bed closely, and at others receded, so that the stream was bordered by sandy banks covered with Cassias, Eremophilas, Grevilleas, gum trees, and Melaleuca. Side streams which had cut out smaller gorges for themselves entered the main stream at intervals and every now and then the bed of the latter held fair sized pools of water, on one of which we counted a flock of sixty-nine teal and duck.

Some nine miles from the mission station, and just where the river takes a big sweep almost due east and west we came upon the first Palm tree and camped for the night in a very picturesque spot where the rocks were broken up into great red blocks piled on one another to form pinnacled masses.

The Palm tree which was first found by Giles, in 1872, is very much like the common cabbage-tree Palm of the eastern coastal district, but this species

(*Livistona Marie*) is peculiar to just the part of the Finke Gorge in which we were camped, and to the Palm Creek which entered it on the west side close to our camp.

Professor Tate and myself spent the afternoon searching along the steep banks of the river at the base of the high cliffs. These banks are formed of the talus of the cliffs, and are covered with a growth of native fig trees and such smaller shrubs as Indigofera. In this part of the gorge there are not more than, at the outside, a dozen mature palm trees, the tallest of which would perhaps reach a height of fifty feet. Many hours were spent by us in search of molluscs, and we were rewarded by the finding of a new Bulimnoid shell in the loose earth and dead leaves under a fig tree on the northern bank of the gorge. There was apparently just this single colony of the shell (*Liparus spenceri*) as, though searching carefully, we never found it except in this one restricted spot. Perhaps if the whole district were searched other colonies would be found, but they are evidently few in number and far isolated from one another, a feature in the distribution of many animals and plants which was constantly being impressed upon us.

Our camp on the soft sand of the creek bed close by a water-hole and at the foot of a small clump of fine gum trees and Palms which stood out against the lofty red cliffs behind them was a very picturesque one. The next morning Professor Tate and myself once more went down the Finke, whilst Dr. Stirling and Mr. Heidenreich rode on up the Palm Creek coming in from the west to see if it were worth our while to go and camp there. They returned after a few hours and reported that it was well worth our going up, so in the afternoon we shifted camp, Dr. Stirling with Messrs. Winnecke and Heidenreich returning to the main camp at Hermannsburg, while Messrs. Tate, Belt and myself went up the Palm Creek intending to spend a day or two there.

After traversing some two miles we came to a part where the hills closed in and formed as usual a big sweep of precipitous red cliffs which rose abruptly from the smooth, rocky bed of the river. The sides of the gorge on the northern bank of the stream were overgrown with Cycads, whilst a solitary palm or two had managed to establish themselves in clefts right in the centre of the rocky bed. Passing out of the Cycad gorge the hills opened out a little, where a stream came in from the south but soon closed in again to form another long, winding gorge leading back amongst the hills.

The river bed was almost entirely formed of smooth rocks, but a little way beyond the Cycad gorge was a patch of sand and on this, as there was unfortunately no chance of a heavy rain to flood the creek, we camped.

Wandering up the gorge we soon came upon the palm trees the total number of which does not exceed if it equals one hundred— that is excluding young seedlings. There is no sand or soil in the gorge the bed of which is completely filled with water during the short time that a flood comes down, and the torrent, judging by the heaps of *débris* piled up against the trunks of the palms, must come down with considerable force and volume.

The rocks are worn quite smooth and amongst them are pools of water three or four feet deep often surrounded by rushes. At each side of the gorge and more especially on the northern under the shade of the rocks is a growth of scrub above which the palms stand out. They are confined as may be seen from the illustration which represents a view along the Palm Creek looking west, to this northern side except a few which are growing right in the bed of the creek. Very young ones only a foot or two in height are numerous in the small clefts amongst the smooth rocks, but there are very few half-grown trees which seem to show that the great majority of the young ones get torn out during flood time and so the colony does not increase in numbers and may perhaps as the older trees die off be actually diminishing. It might have been expected that the floods would have washed the hard fruits away to other parts of the river where they would have germinated. Probably the few trees along the main river have been thus transported, but they are very few in number, so that it appears as if this method of spreading the species were of little avail and that like many other species the Palm exists only in a very restricted area. The reason why the Palm seeds do not germinate freely when carried, as they must be, down into the main Finke is possibly due to the fact that the drying up waters along the latter are frequently brackish in nature and so perhaps the vitality of the seeds may be impaired. Whilst plenty of very young seedlings were to be found along the Palm Creek scarcely one was seen along the main channel of the Finke.

Along the gorge young Cycads (*Encephalartos Macdonnelli*) were also found, but the adult plants, apart from those on the cliff sides already referred to, were few in number and were principally seen along the cliffs bounding one or two side streams which entered the main gorge.

The tallest Palm was fully eighty feet in height and one or two of them had curious cork-screw trunks. It appeared to be rather like sacrilege to touch the trees, but as we were anxious to find out if the leaf sheaths harboured any special forms of life, one, about sixty feet in height, was cut down. After carefully removing every leaf the only animals found were a solitary cockroach and a bug; there was no trace of anything like a mollusc or a planarian worm sheltering under the

broad sheathing leaf stalks—nor it may be added was there any trace of animal life save an odd mollusc and an insect or two amongst the Cycads which were carefully searched. A view of the Palm Creek, such as the one figured, with a rock-pool in the foreground and the Palms rising above the scrub gives one almost the idea of a semi-tropical scene, but in reality there was none of the damp luxuriance which we had hoped might perhaps be met with in this spot. Away from the margin of the water-pools everything was as dry as usual, but as we were anxious to examine the flora and fauna of the ranges more minutely than we had yet been able to do we determined to camp here, as it did not seem likely that we should find a more favourable spot.

Accordingly we sent our black boy back with a camel to the main camp, as previously arranged with Mr. Winnecke, for a supply of food, and then Messrs. Tate, Belt and myself spent three days searching up and down the creek itself, on the cliffs bordering it and up the side streams flowing into it.

It was our longest spell in one camp and our collections were considerably enhanced by the chance which it gave us of a more thorough examination of one spot than we had been able hitherto to make, especially as regards smaller forms such as molluscs and insects, while at Hermannsburg the stay enabled Mr. Keartland to add largely to the collection of birds.

A short account of the animal life of this spot will really serve to describe that which is generally met with around any of the water-holes amongst the ranges.

In the water-holes there were at least six species of fish, none of which were as large as the bigger ones caught in the Redbank Creek. They were *Therapon trutaceus*, *T. percoides*, *Nematocentris tatei*, *N. winneckei*, *Eleotris larapintae* and *Chatoessus horni*. The water-holes were all isolated from one another and had rocky beds with but little sand ; the smaller ones would soon dry up, but one or two of the larger ones which were some twenty yards long would probably persist for some length of time, though there were apparently none which would last through anything like a drought—not even a short one—as there was no constant supply of any kind such as exists in the sandy pools such as the one at Henbury, where the water is forced to rise to the surface after flowing along beneath the sand.

Of molluscs the pools contained six species—*Melania balonnensis*, *Limnaea vinosa*, *Bulinus texturatus*, *B. dispar*, *Planorbis fragilis* and *Ancylus australicus*.

Most of these as might have been expected are widely distributed through the Larapintine region, but of one species, *Isidorella newcombi*, which is otherwise widely distributed, we did not obtain any specimens.

With the land molluscs, which have no such means of distribution as the fresh water ones, the case is very different. As Professor Tate has pointed out, and as was frequently impressed upon us whilst collecting, they occur in very contracted areas, sometimes as already noted in, for example, the cases of *Angasella arcigerens* and *Liparus spenceri* we only found single colonies. Out of a total of twenty-five species secured during the expedition fourteen were found in and about Palm Creek, of which four, viz., *Endodonta planorbulina*, *Chloritis squamulosa*, *Liparus spenceri* and *Pupa ficulnea*, were found nowhere else. This remarkable restriction of species to small areas is very striking and is best exemplified amongst animals by the mollusca which, while they can persist in sheltered spots, have very little opportunity owing to climatic conditions of wandering far away from their hiding places and of establishing themselves elsewhere and so perhaps for long years a single colony will occupy a spot it may be only a few yards square. They are in addition very liable to extinction, as when the rain falls it washes in torrents down the cliff sides on which they shelter, and as shown by the number of dead shells in the rejectamenta of the river all along its course numbers must get carried away and perish; just those which are in the cracks and crevices and most sheltered spots alone being preserved.

The pools were all full of clear water so that as usual the Estherias were dead, but that they exist during the time when the water is muddy was shown by the presence of their dried carapaces in one or two spots where the pools had dried up. There was only a single species *(E. dictyon)* to be found; its carapace was slightly more than an eighth of an inch in length, and when magnified was seen to have a raised network pattern between the lines of growth which had the appearance of cells of a honeycomb cut across. It is a new species and was only met with in this one spot, but in the South Australian Museum are specimens of the same species the exact locality of which is not known, though they serve to show that it does exist elsewhere.

The Estherias are more characteristic of the pools on loamy flats as they prefer muddy to clear water, in fact as before said they do not seem able to survive in the latter, and as along the Palm Creek there is very little sand or loam and the water lies in clear rock pools the surroundings are scarcely suitable for them. Around the edges of the pools little Hylas *(H. rubella)* were found belonging to the one widely spread species and, where there was sand, there the blacks without difficulty obtained specimens of the burrowing frog, *Limnodynastes ornatus*.

Lizards were not especially abundant, only eleven species being found. Amongst the Geckonidæ species of *Heteronota* and *Gehyra* were found, and amongst the Pygopodidæ the thin bodied and elongate *Lialis burtoni*. The Agamidæ, so numerous in the sandy and loamy plains, were only represented by the widely spread *Amphibolurus reticulatus*, whilst the most abundant forms belonged to the family Scincidæ which was represented by such widely spread forms as *Egernia whitii* and *Hinulia lesueurii*. In addition to these two, three species of skinks, viz., *Rhodona bipes* and *Ablepharus greyi* and *burtoni* were found here and nowhere else, and a single specimen was seen of a new red-tailed variety of *Ablepharus lineo-ocellatus* which is somewhat widely spread, being found from Alice Springs in the north to the Goyder River in the south.

A considerable amount of time was spent in collecting insects of various kinds :—under stones by the water-side there was of course the common little carab *Tachys spenceri* and two species of Staphylinidæ, a small black and a larger black and red one (*Philonthus subcingulatus* and *Cryptobius mastersi*). Turn up any stone by the side of a water-hole in the James and McDonnell Ranges and you will be sure to find the first and at least one of the latter two.

The flowering shrubs were as usual disappointing. *Cassia artemesioides* was covered with masses of bright yellow blossom, but scarcely a single insect was to be secured by shaking except certain Curculionidæ which were everywhere more in evidence than any other form of beetles. I never saw an insect in the Cassia flowers, and it was a curious fact that though everywhere the shrubs were flowering luxuriantly the pods formed were but few in number and most of them contained only ill-formed seeds. If the flowers be entomophilus as is most probably the case then they were evidently suffering from a lack of insect life which is probably to be associated with the low temperature at night-time, and the frequent occurrence now that it was (July) mid-winter of a biting south-east wind during the day-time.

To secure beetles in Central Australia you really want to be there during the rainy season—in fact during a succession of seasons for just before and just after a heavy rainfall they appear for a short time and then rapidly disappear.

However, during our three days' spell at Palm Creek forty-seven species were collected of which twenty-five were new. To anyone who has had the experience of collecting in Central Australia during the dry months when at night your water bags are frozen solid this will not probably appear so small a number as it will to those whose collecting has been done in more favourable spots. During the whole Expedition we secured one hundred and seventy-seven species of which sixty six are new. The large proportion obtained during three days in one spot was due to the fact that we had one of our rare spells and were able to do a little more careful searching, but, judging by the way in which throughout the ranges we came upon the same animal time after time, I do not think it very likely that the spending of a longer time in other parts would, at this particular season of the year, have added *proportionately* to the collection of Coleoptera though of course it would have added a certain number of new forms.

The same or any other spot would undoubtedly yield different species at a different season or time, and so far as collecting insects of all kinds is concerned it must be remembered that our work was done during the most unfavourable season.

Amongst the Arachnida the more common forms were a species of Myriapod resembling a Scolopendra and the Cermatia (Scutigera), which appears to be identical with the one commonly found in other parts of Australia such as Gippsland in Victoria. In some parts of the world, as for example in Malta,* they are described as coming out into the blazing sun in search of their prey. I have collected a considerable number of Myriapods in various parts of Australia, such as Victoria, Tasmania, Queensland and the Central district, but have never yet seen a Scutigera out in the open. They always lie under logs or stones or the bark of trees and when disturbed always move away with remarkable speed into a dark spot. Their legs move in such a way and so rapidly that you can only see as it were a series of waves passing down each side of the body, and there is in Australia at all events no Myriapod which in speed of movement is to be compared with the Cermatia.

Sometimes they are beautifully coloured with the flattened, plate-like terga blue and red and their gliding movement is so rapid that it is no easy matter to catch them without the loss of a few legs, which come off almost with the slightest touch.

The Scorpions at Palm Creek were found under stones. These creatures adapt themselves to the nature of the country in which they happen to live. If it be a rocky spot then they live under stones, if it be amongst the sandhills then they burrow. At Crown Point for example, in sandhill country you could, during

* Peripatus, Myriapods and Insects, Pt. I., Camb. Nat. Hist., p. 35.

the day time, only secure a scorpion by digging it out of its burrow. The latter can easily be distinguished by the marks at the entrance made by the legs of the animal. There is a small hole on the surface with a little flattened-out heap of sand marked all over by curious and very characteristic lines as if the blunt edge of a knife had been pressed down on the sand in such a way that all the short depressions thus made in the sand converged towards the mouth of the burrow. To find the animal you may have to go down four or five feet till you come to a small chamber in which it lies at rest during the day time.

Amongst other Arachnids the most interesting were two species of, or allied to, Chelifer, one under the bark of a tea tree (Melaleuca sp.), the other amongst the *débris* under a fig tree.

Spiders were fairly plentiful; we secured in this part eleven species, of which the most abundant was a species of Isopeda which here as elsewhere was to be found under the shelter of stones close to the edge of a water-hole. Probably this species feeds on the small beetles belonging principally to the Carabidae and Staphilinidae found in the same situation. Amongst plants the two most striking forms, the Cycad and the Palm, have already been alluded to. In the water-holes *Naias major* (the fruit of which was obtained only here), *Potamogeton Tepperi* and *Triglochin calcitrapa* were growing, and around some of them were thick beds of bulrushes and reeds. The saxatile plants were those characteristic of the region; fig trees of two species (*Ficus platypoda* and *F. orbicularis*), and growing amongst the figs and sheltered and supported by them was a species of native orange (*Capparis spinosa*), very different in the nature of its lax growth from its more common ally *C. Mitchelli*, which forms at times a small tree on the rocky hill sides. Here and there were odd shrubs of the red-flowering *Grevillea agrifolia*, with a mistleto (*Loranthus gibberulus*) parasitic upon it. Patches of *Cassia venusta* with yellow blossom, and of Eremophilas with pink and lavender bloom grew on little flats often high up the rocky sides, and here and there the large yellow flowers of *Hibbertia glaberrima* stood out, often forming the only bit of bright colour in a shady gorge.

The characteristic plants of the Upper Steppes are to be found amongst what Professor Tate has described as the saxatile species, that is those growing on the sides of the gorges, on the basal part of the escarpments of the hills and on the talus which slopes down from the escarpment to the valley below. Out of seventy species of flowering plants found growing on and restricted to the rocks no fewer than sixty-three are endemic. The saxatile flora of this region has in recent years been found to extend to outlying ranges such as Mount Olga, and the

Musgrave and Everard Ranges in South Australia, and on the Cavenagh Range in West Australia, so that botanically these may be regarded as outliers of the Higher Steppes.

After three days hard work in the Palm Creek we reluctantly returned to the main camp as it was necessary for us to push on towards Alice Springs. Early in the afternoon we reached Hermannsburg, and the rest of the day was spent in packing up and labelling all the specimens secured. This always occupies a considerable amount of time, and must be done carefully if the animals are not to be spoilt. Each fish for example must be separately wrapped up in calico or muslin or else the fins get broken and the scales rubbed off.

At Hermannsburg Mr. Keartland had been hard at work amongst the birds. The most important addition made here to the collection was a new species of Xerophila (*X. nigricincta*) which is distinguished from its close ally, *X. pectoralis*, found in the Port Augusta district, by the presence of a black instead of a cinnamon-brown band across the chest. Amongst the scrub were a large number of the, popularly called, superb warblers and as usual the dull-coloured females were far more numerous than their richly-coloured mates, who kept out of sight as much as possible as if they were quite aware that their brilliant colouration would make them too conspicuous objects to their enemies for them to be safe in the open. Their rich, sapphire and cobalt-blue colour set off with patches of brown and white and jet black bands render them perhaps the most beautiful of all the birds seen in the scrub. There were three different species to be seen, *Malurus melanotus*, *M. lamberti* and *M. leucopterus*, of which little flocks of the two latter were often found feeding sociably on the same bush.

Here and elsewhere the "cat bird" (*Pomatostomus rubeculus*) attracted attention to itself. Its call is a most peculiar one—just like the mewing of a cat—and the birds which are very sociable are constantly uttering their cries as they jump from branch to branch and perform most curious antics. At Henbury Mr. Keartland watched three birds carrying wool from an old sheep skin to a nest whilst a fourth was engaged in arranging it, so that very likely two pairs may share a single nest. This bird was one of the comparatively few ranging southwards into the Central district from North and North-west Australia.

The black boys had caught a good number of a common mouse which Mr. Waite has described as new under the name of *Mus hermannsburgensis*. It appears to be very common in the Missionary Plains and along with it they had also captured two male specimens of *Antechinomys laniger* which is one of the rarest and most difficult to secure of the smaller marsupials. At first sight in shape, size and colour when hopping along on the ground it bears a striking resemblance to the little rodent *Hapalotis mitchelli* which is widely distributed over the central district but it is more slender in build and of course the shape of the head, when seen close to, distinguishes it at once. The most numerous of the smaller mammals are undoubtedly the various species of Mus and Hapalotis, next to them but far less common is probably *Sminthopsis crassicaudata*. The scarcity of Antechinomys is rather strange as its habits do not bring it into direct competition with the rodents except so far as each of them has taken on the same method of travelling by jumping. All the small rodents and the marsupials referred to live side by side in burrows on the hard loamy flats amongst the scrub, and in the matter of speed the marsupials, so far as can be judged, can get over the ground as rapidly as the rodents.

It is, however, quite possible that the female when carrying young is somewhat handicapped and may be more easily caught by birds of prey and, as noted elsewhere, a very slight difference in speed when a hawk is in pursuit and the little animal is seeking the shelter of a bush or tussock of grass may save, or lose it, its life.

On Monday, 9th July, we left Hermannsburg and travelled eastwards over the Missionary Plains. At noon we struck the Ellery Creek bordered with good sized red-gums and containing along its bed a few scattered water-holes: a few miles to the south of us it ran into a gorge in the James Range on its way to join the Finke to the south of the Glen of Palms.

The plain was slightly undulating and covered with the usual scrub of Mallee gum (principally *Eucalyptus oleosa*), Mulga, Cassias and Eremophilas. There were now and again very fine specimens, as much as forty or fifty feet in height, of *Acacia salicina* the leaves of which in some cases hung down below the twigs leaving these bare above, so much so that the tree had sometimes the appearance of a weeping willow. In other cases the pendant arrangement was nothing like so strongly marked. We saw also in various parts curious modifications of the Mulga; its foliage varied considerably in hue from an olive-green to bluish-grey. Amongst the sandhills, for example, between Lake Amadeus and Ayers Rock the latter tint prevailed and in addition the thin branches were given off almost horizontally from a central stem forming a tree of a very different appearance from that seen in most parts where the branches were nothing like so horizontally disposed but more divergent like the ribs of a fan and the foliage was more olive-green in colour.

In the accompanying illustration two Mulga trees are shown, the Mulga scrub to which such frequent reference is made in all descriptions of the interior of Australia consists of a dense growth of trees such as these. Their thin, wiry branches, when dead, are like long thorns and are very apt to run into the feet of horses or camels and frequently give rise to painful, festering sores. As a general rule the trees, which do not usually reach a much greater height than fifteen or twenty feet and often less, grow very close together—so close that, next perhaps to travelling over Porcupine covered sandhills, the penetration of Mulga scrub is the most disagreeable and disheartening task attendant upon journeying through Central Australia.

MULGA TREES (*Acacia aneura*).

We camped in the scrub after travelling some twenty miles. Our time was rapidly drawing to a close as we were really due back in Adelaide at the beginning of August and it was now 9th July and we were still some little distance from Alice Springs, the journey down from which even though it lay along the overland track would occupy some three weeks.

Mr. Horn's rough sketch of the route which he desired us if possible to follow indicated our striking somewhat northwards again so as to reach the McDonnell Ranges at or about Paisley Bluff. Mr. Winnecke's previous experience showed that there would be very great difficulty attending upon any attempt to take the main camel train across the ranges, in fact that it was out of the question to try to do so. It was therefore decided that the main train under charge of Dr. Stirling should continue travelling eastwards along the Missionary Plains and then reach Alice Springs by way of Owens Springs, always keeping to the south of the

McDonnells. A small party consisting of Messrs. Tate, Belt and myself with Mr. Winnecke was to make for Paisley Bluff and there find some way right through the ranges to the Burt Plain and then travel eastwards to Alice Springs.

Next morning we started off travelling north-west across the undulating plain with every now and then patches of travertine and stony flats. In the afternoon as we got nearer to the range we saw a long series of low, jumbled hills, above and behind which rose high peaks which were evidently Paisley and Brinkley Bluffs.

Kangaroos, red males and grey females, and Bettongias, were fairly abundant and there were plenty of large, wedge-tailed eagles flying about and perching on the trees close to us, some of them very light, others very dark brown in colour. The Mulga scrub got thicker and what with this and the Porcupine covered hills it was rather rough travelling. Just at sunset we got into a regular jumble of the rounded Porcupine grass hills, which, as before, looked quite smooth and beautifully down-like in the distance. After following up a small valley into the hills Mr. Winnecke luckily came across a small spring issuing from the conglomerate rock of which the hill was formed. As the previous night had been spent at a waterless camp we were glad to give the two horses which we had with us a drink. Where it issued from the ground the water was quite warm, but it only formed one or two very small pools each about a yard long and an inch or two deep and then disappeared.

We camped amongst the tea-tree by the side of a dry creek, and in the morning sending the camels and horses round the base of the hill we climbed up to the top above the spring to get a general idea of the country. As far as we could see the Missionary Plains stretched away to the west their flat surface broken only in one spot where the solitary Gosse Range stood out. To the south was the northern part of the James Range known as the Waterhouse which was continuous at its western end with the Krichauff Range, while eastwards it trended towards the north so as gradually to narrow in the plain between it and the McDonnells on the southern ridge of which we were standing.

Across the plain to the east of us a streak of gum trees marked the course of the Hugh River which ran right into the Waterhouse Range. It appeared to end abruptly against the latter, but in reality it passes as usual right through it in a gorge at Owen Springs. We were standing on the hills bounding the Horn Valley to the south and to the north of us east and west stretched the McDonnell Range in which Paisley and Brinkley Bluff and Mount Conway stood out conspicuously. Climbing down the hill we joined the camels and soon came out into a level Mulga flat—the Horn Valley—and then, following up a branch of the Hugh, passed through the ridge bounding the valley on the north, the ridges being cut through here by one or two dry gorges. Emerging from one of these narrow gorges we found ourselves on a plain perhaps three-quarters of a mile wide running east and west with masses of gneissic rock projecting here and there. To the north of us lay the McDonnells proper and the plain along which we travelled eastward was evidently continuous with that which lay at the base of Mount Sonder and on which we had previously camped at the junction of the Davenport and Redbank Creek.

After about two miles easy traverse we came to the Hugh River and halted for an hour by the side of a small water-hole just close to where the Hugh runs in a deep gorge through the ridge on the north of the Horn Valley.

We were rather surprised to find camel tracks—evidently recent ones in the sand by the side of the water-hole. They could only have been made by Mr. Watt's party, as no one else was likely to be travelling in the district, and our black boy said that they were only a day or two old.

Our difficulties now began as we wanted to find some way in which to pass through the ranges to the north of us. After a short halt we travelled eastwards hoping to find a way round the base of Mount Conway, but finding this impracticable we retraced our steps and followed the Hugh. The country was rough and rocky and by no means easy work for camels, but after some few miles we came to a good water-hole, and late in the afternoon camped in a most picturesque spot just to the south of Paisley Bluff. This water-hole was in a gap in a range flanking the main one.

The main branch of the Hugh ran eastwards from our camp for half a mile in the valley between us and the high ridge, from which rose in front of us Paisley Bluff, and to the north-east Brinkley Bluff. We made our camp, sleeping as usual in the open on the soft sand of the creek bed. In the gum trees the "mopokes" (*Ninox boobook*) were calling to one another, and as it was bright moonlight we could see the dingos sneaking round our camp, but our presence and camp fire evidently prevented them from coming to water.

As this would be our last chance of collecting in the ranges we determined to spell for a day. Early next morning we were out collecting, and followed up a branch creek to the base of Paisley Bluff. It ran through a narrow gorge at the western base of the Bluff. The bed was strewn with rocks of various sizes, amongst which were a few very small and shallow water-pools. White stemmed gum trees (*E. terminalis*), a large species of Melaleuca, with papery bark and reaching a height of forty feet, and shrubs such as *Cassia glutinosa* with its yellow flowers, and *Grevillea agrifolia* with clusters of red blossom were growing amongst the rocks and filling up the small space left between the precipitous cliffs, the sides of which were studded with pines and cycads.

The animal life was just the same as that to which we had grown accustomed around the water-holes amongst the ranges. A few species of beetles and myriapods, and the little frog *Hyla rubella* were abundant under the stones close to the water. On the hill sides rock wallabies were numerous, but there was the same disappointing absence of anything like a rich and varied fauna. Stones could be turned up, flowering shrubs shaken, and bark stripped off trees hour after hour without finding anything to reward one's labour except perhaps a new mollusc sheltering in the *débris* beneath the fig trees, or hiding in crevices amongst the stones. I gave up finally all idea of finding any such thing as *Peripatus*, or a land planarian, or anything more than a stray earthworm in a country where it may be for months together the only moist place lies actually in a water-hole.

It was only the quiet accumulation of specimens gathered day after day which resulted in the finding of as many forms of animal life as we did but the total yield was in no degree commensurate with the amount of time spent in obtaining it, and the most galling thought was that just a day or two's rain would bring out from their secure hiding places so many animals of whose existence not a trace was now to be seen. At the same time I should be much surprised if even after rain such soft-bodied animals as land planarians or slugs were to be found as the class of country is pre-eminently unsuited to them.

We could not help being struck with the dominance of particular forms amongst both animals and plants. Amongst the former of course ants were the most notable, but in addition to these which were found under every stone or log — I doubt if ever we turned up one without finding an ant except such as were right at the water's edge and even here they were sometimes to be seen — there were other dominant forms such as certain species of Carabidæ and more especially of Curculionidæ. At the time of our visit the latter was, apart from flies and ants, in point of number of specimens by far the most extensively represented family of insects whilst amongst the amphibia the little *Hyla rubella* was found at every water-hole from the Adminga Creek in the south to Alice Springs in the north, and westwards right throughout the ranges.

Amongst plants certain genera were equally dominant. On the flats and along the valleys amongst the ranges Cassias, Acacias (especially *A. aneura*), Eremophilas and Eucalypts formed the mass of the vegetation and on the rocks the Pine tree, whilst the Porcupine grass (Triodia sp.) dominated alike both valleys and the rockiest hill sides.

Under a block of quartzite in the bed of the gorge I came across a small nest of honey ants of which numerous specimens have since been found by Mr. Cowle. It was a very different form of nest from that of *Camponotus inflatus* but this was probably only a young colony. Burrows branched off in all directions but did not go far down. The ants were of a rich golden colour and the insects were nothing like so swollen out as in those of the first-named species. Though their abdomens were inflated so that the terga and sterna were all separated from one another still they were capable of a certain amount of movement.

In the Mulga scrub at the base of the range there were two forms of ant nests which were frequently met with everywhere amongst the scrub from Ayers Rock in the south to the Burt Plains in the north. One has the form of a mound upwards of two feet in diameter and about six inches high, with a large, crater-like depression at the top. Around the sides of the mound the ants arrange a thick deposit of dead Mulga leaves all placed radially in a perfectly regular manner. The other mound is the same size but instead of the crater depression it has a slit from three to six inches in length and half an inch to an inch in width and is always covered over with various kinds of dried grass seeds; the nest is inhabited by one of the numerous species of Camponotus (*C. denticulatus*). Both of them have passages leading away in various directions, but though I spent some time here and elsewhere in trying to follow them up the hard, stony ground prevented this being done satisfactorily and the large black ants inhabiting them, which were from half to three-quarters of an inch in length, enforced a certain amount of carefulness as they naturally objected to having their homes broken up.

It is difficult to see what is the use of the Mulga leaves and the grass seeds as I could detect nothing such as a fungoid growth amongst them, though this might be present under different climatic conditions, or anything which could be of service to the ants, and whilst the Mulga leaves might serve to drain off water during the rain season the grass seeds would rather have the opposite effect.

After spending the day collecting in the gorge and along the flats by the creek and on the hill side we went back to camp and found that the black boy had brought in five rock wallabies (*Petrogale lateralis*). This is at once distinguishable by the light line along either side of its body and though usually spoken of as the West Australian rock wallaby it is widely distributed over the Centre, occurring on the Desert Sandstone ranges and throughout the George Gill, Levi, James and McDonnell Ranges, in fact it is probably to be found amongst all the hill country of the Central area.

The average length of the body is two feet and the tail is just the same length as the body. Three of the specimens were females and each of them had a single young one in the pouch, so that probably this may be regarded as the usual number produced at each birth. The young ones grow to a considerable size before leaving the pouch, and as the rock wallaby lives exclusively amongst the hills, never apparently spending any time in the flats, a large number of young ones to be carried about at a time would be a serious handicap in a region where birds of prey such as the wedge-tailed eagle are constantly on the look out for food.

The explanation of the way in which such an animal as the rat or the rabbit if introduced into a region previously occupied by marsupials soon exceeds in number the lower forms is probably closely connected with this manner of carrying the young.

In the first place, at an age when a young marsupial at sight of danger at once flies to its mother's pouch a young rat or rabbit is taking care of itself. If a hawk or eagle catches the mother rabbit the young one is left or *vice versa*. In the case of a marsupial the mother has to carry the young ones, and not only does the extra weight prevent her gaining shelter but, if caught, both she and the young ones are sacrificed. As already pointed out, a very slight difference in speed will save or lose the animal its life. When hard pressed a kangaroo will throw the young out of the pouch so as to be able to travel faster. In fact this habit of carrying the young one for so long in the pouch is a severe handicap for a marsupial when it comes in contact with a rodent, for though they may not compete with one another directly so far as their food supply is concerned - though many of them do this—still they both have to avoid a common enemy in the nature of birds of prey. In the case of such smaller marsupials as, for example, species of Sminthopsis in which the number of young produced at a birth is from eight to ten and there are at least two broods in each year it is a matter of considerable surprise that they are not much more numerous than they are. The explanation is probably associated with the fact that there is a considerable length of time during which not only does the capture of the mother result in her destruction and in that of all the young ones, but that during this period she is severely handicapped by not being able to reach shelter rapidly. It may perhaps be objected to this that such an animal as a rabbit is handicapped by having to carry the young ones in utero for a much longer time than the marsupial does, but anyone who has seen the well-developed, pouch young ones of a marsupial will realise how much more cumbersome a burden they are than the uterine embryos of such an animal as a wild rabbit.

Early on the morning of 13th July we left camp intending if possible to get through the ranges and camp the next night on the Burt Plains. Leaving Paisley Bluff to the west we followed up the Hugh until we came to the gorge, through which it has cut a way for itself just at the eastern base of Brinkley Bluff. This spot is interesting, as it was through this very gorge that in March, 1860, McDouall Stuart was able to make his way across the McDonnells and to reach for the first time the centre of the continent. The creek bed in the gorge was occupied by a water-hole leaving just enough room for the camels to pass. North of the gorge we found ourselves in a jumble of low hills covered with Porcupine grass, Eremophilas, Cassias and Acacias, and at noon halted by a water-hole to give the camels and horses a rest.

To the south of us the main range could be seen stretching east and west with Brinkley Bluff standing out clearly ; to the north nothing but low rough hills could be seen. About five or six miles north of the range we crossed the watershed, and from this onwards the small creeks flowed northwards.

For some hours we were winding in and out and over the hills—very difficult travelling for the camels. Just at sunset we led them up a high gneissic range and with considerable difficulty, as ugly rocky ledges had to be climbed, we reached the top and saw stretching far away to the northern horizon the broad, scrub-covered Burt Plains. To the north-west lay Mount Solitaire, and away in the distance isolated hills could be seen, whilst eastwards the McDonnell Range trended somewhat towards the north.

It required considerable care to take the camels safely down the steep face of the hill, but at length we reached the plain and camped at dusk in the sandy bed of a dry creek.

A flock of more than fifty black cockatoos were screeching overhead evidently much disturbed by our appearance on the scene. We had only travelled in a direct line some sixteen miles, but the country had been so rough and difficult that it had taken us ten hours' hard work in which to traverse this short distance.

At Brinkley Bluff we had been surprised to see the tracks of Mr. Watt's party returning southwards, so it was evident that he had not been able to make his way across the hills to the Burt Plains. We learned subsequently that he had attempted to cross more to the westward, but as he and the two prospectors with him were travelling with only one baggage camel to carry provisions the likelihood of striking the Burt Plain at a long distance from any water supply had very wisely caused them to turn south again into the ranges. They had followed down the Hugh under Brinkley Bluff to the water-hole by which we had first seen their tracks and then had struck eastward to the south of the main range and so had reached Alice Springs.

Our camp on the plain was at a height of 2185 feet and the night as usual was very cold. All the next day we travelled eastwards along the base of the hills through the open scrub. The ground was covered with dried up yellow grass and the scrub of Mulga, Cassias, Santalum and gum trees was as monotonous as usual. Every now and again a small gum creek ran out for a short distance away from the hills, but everything was perfectly dry except at one spot (Painta Springs) where there was a small soakage with one or two small water-pools in which we secured a few golden-spotted water beetles.

A well has been sunk here by the side of which a large date palm is growing and the relics of feeding troughs show that it has once been used as a watering place for one of the outlying cattle runs. This spot and the Missionary Station at Hermannsburg were the only two at which we saw the Date Palm, though a considerable number of seeds have been planted by different explorers in what appeared to them to be suitable spots.

At one spot we came across a small patch of the mound nests of what are called the meridian or compass ants. These are found in other parts of Australia such as near Cape York and Port Darwin and the curious feature about them is that the mound, which is three or four or even five feet high, is flattened from side to side in such a way that the broad sides face east and west, and the narrow ends north and south. As it tapers upwards it has, seen from the north or south, a wedge shape. There were altogether perhaps a hundred of these occupying half an acre of ground and their shape and bright red colour render them very striking objects. Unfortunately we met with them in the middle of a long march when it was quite impossible to stop and examine them and my hope that we should afterwards meet with others in similar country was not realized. They are made and occupied by a species of Termite or white ant and the only other white ant mounds which we saw were a few small, grey-coloured ones about eighteen inches high on some flats near Lake Amadeus.

At about twenty-five miles from our last camp we once more struck the overland telegraph line and the track which runs straight across Australia from Port Darwin in the north to Adelaide in the south. There is no difficulty in following this and after going on for about two miles to the south we camped and in the morning reached Alice Springs. The Telegraph Station lies in a picturesque spot just to the north of the main McDonnell Range. After halting for a few minutes at the station we went on along the Todd River and through the small township of Stuart to the Heavitree Gap on the south side of which we found the camel train camped. In the evening we retraced our steps to the Telegraph Station where we were made welcome by Mr. and Mrs. Gillen and, through the kindness of Sir Charles Todd, we were enabled to communicate by telegraph with our friends in Adelaide, Melbourne and Sydney. After a rest of three days which were utilized in collecting round Alice Springs, the main party travelled southwards following the well beaten track to Oodnadatta, which was reached early in August.

At Alice Springs the expedition practically came to an end, but Mr. Watt and myself stayed behind, the former to pay a flying visit to the gold and so-called ruby fields, whilst my own time though fully occupied with work of various kinds was spent more pleasantly as the guest of Mr. and Mrs. Gillen at the station. Mr. Gillen kindly sent blacks out in search of animals which I was especially anxious to secure and to make sketches of and colour notes with regard to, in their living state, as our travelling had been so hurried that there had been little chance of doing this whilst we were on the march. I was also especially anxious to secure if possible some more specimens of a small new marsupial (*Phascologale macdonnellensis*), and to watch the so-called "barking spider" in its natural state. Of the former only a single specimen had been obtained, and this was a male. Fortunately the blacks caught two more whilst I was there, both of them females. They are very active little creatures the size of a small rat but with a great swollen tail which is strongly incrassated. They live amongst the big blocks of rock on the hill side, and so are very difficult to secure especially in the dry winter months when they do not come out. The offer of a shirt and a lot of tobacco failed to secure more than two, though since we returned I have received several more, thanks to Mr. Field and Mr. Gillen.

This curiously fat tail is seen not only in this marsupial but also in *Phascologale cristicauda* and *Sminthopsis larapinta*, and an examination of the tail of each of these species shows that the swelling is due to the deposition of a great amount of fatty material ; in fact when the skin is cut through, the back bone is found to be embedded in a surrounding swollen mass made up of fat and yellow elastic tissue. The tail seems to be equally swollen out at all seasons of the year.

A rodent captured at Alice Springs and described by Mr. Waite under the name of *Conilurus (Hapalotis) pedunculatus* has also a somewhat swollen tail, but in this case it is also brittle and pieces of it snap off easily when handled : possibly, as Mr. Waite suggests, the breaking off may, as in the case of lizards, be useful in aiding the animal to avoid capture by the loss of part or all of its tail, but in the specimens yet secured, though it is the exception to find one with a perfect tail, there is no appearance of any fresh growth indicating that the lost part can be replaced by subsequent growth.

At the Alice Springs there is still the same general arrangement of the ranges as was met with elsewhere, but owing to the general width of the valley through which the Todd flows there is no difficulty except when the narrow Heavitree Gap is filled with water in traversing the ranges from north to south. At this point the McDonnells are about twenty miles in width. The low jumble of hills flanking the main range on the north are from ten to twenty miles wide. At the southern end of these lies the Telegraph Station by the side of the Todd, which is here of considerable width, but as usual in the dry winter months its sandy bed contains only a few pools of water. The station is built on a high bank by the side of a pool which lies at the base of a projecting rocky cliff, on which grow fig trees, and on the southern aspect the very pretty white flowering Plumbago (*P. Zeilanica*).

About a mile and a half to the south of the station the main ridge of the McDonnells is crossed. Just at this point its height is insignificant, but away to the west high peaks such as Mount Conway can be seen arising from it. Then follows a broad flat valley on which is built the little township of Stuart. It consists of a few stores and the inevitable hotel : camel teams not infrequently pass up and down the telegraph line bringing stores to the township, telegraph stations and outlying cattle runs, and one or more of them are often to be seen camped outside the township in the scrub.

This valley is a continuation of the one which lies at the base of Mount Sonder. To the south it is bounded by a high ridge—the most prominent feature in this part—of which the highest point is Mount Gillen, the top of the escarpment of which is fully 3,000 feet in height. A bold precipice, facing north, three or four hundred feet high is succeeded by a steep talus.

Through this ridge the Todd breaks in a fine gorge known as Heavitree Gap ; on the sandy bed, as is usual in the larger gorges, the red gum grows and a few cycads are dotted about on the precipitous cliffs. To the south of the gorge comes another wide valley, the eastern continuation of the Horn Valley. In this the police camp is placed and here, close to the Heavitree, was our camp. Then came another range cut through by another gap and then a broad valley, the continuation of the Missionary Plains.

Within a radius of twelve or fifteen miles of the Alice Springs station are some of the most picturesque gorges to be found amongst the ranges. To the east lies Emily Gap, some twenty or thirty yards wide, completely closed by a deep water-pool; westwards from this we come to the Heavitree Gap and then about twelve miles still further west is Temple Bar Gap with a broad sandy and gum tree covered bed, and still further west again is the gorge through which the Jay flows south to join the Hugh River. All these lie in the ridge forming the northern boundary of the Horn Valley. Just to the north of Temple Bar, only cutting through the main McDonnells, is Simpson's Gap, perhaps the most picturesque of all, with its rugged precipitous red rocks rising abruptly on either side of a deep water-pool not more than fifteen feet in width.

Shortly after our arrival at Alice Springs we had been shown by Mr. P. Squire the empty carapaces of a large Phyllopod animal. It looked as if it belonged to a very large flattened Estheria nearly an inch in length. Accordingly, under the guidance of Mr. Field, one of the Telegraph Station staff, Professor Tate and myself went out to a big clay-pan known as Conlin Lagoon. Our way lay along the Todd to the south of the ranges and then we struck along the easterly continuation of the Missionary Plains, here not more than a mile broad.

Turning west we passed the racecourse, the scene of considerable excitement at Christmas time when the annual meeting is held, attended by all of the scattered inhabitants of the central district for several hundred miles around. The Grand Stand, made of planks and brushwood, looked, being out of the season, somewhat dilapidated, and the lawn and flat were occupied by a hard, sun-baked and cracked expanse of dried mud, the course being indicated by a wide circle of posts at intervals. Between the course and the township of Stuart, which during the racing carnival is crowded, lies the Heavitree Gap, through which all traffic must take place, and if one of the summer downpours happens to occur suddenly during a race day, then to the excitement of the racing is added that of the chance of a flood coming quickly down the gap and cutting off the retreat to the township. In Central Australia a river bed quite dry in the morning and hard to traverse by reason of its thick soft sand may in a few hours be transformed into a roaring torrent. However when we passed it the gap was quite dry and the Grand Stand, lawn and flat deserted.

Some few miles to the east we came upon the lagoon which is in reality only a clay pan. At the time of our visit, that is during the dry season, it was still of considerable size being about a quarter of a mile wide and three-quarters of a mile in length, but the indications of flood on the surrounding flats show that in the rain season it must be of considerable extent. It is simply a shallow depression between the two ranges—not more than at most five feet deep with a clay-sand bed serving to retain for a time the water which drains into it as there is no outlet either east or west. Water beetles were darting up and down in the muddy water and in the main lagoon *Estheria packardi* with its blood-red appendages was to be seen but not a trace of the larger form (*Limnadopsis squirei*) alive. Even the empty carapaces were quite confined to the dried up and scrub covered flats to the south and east of the lagoon, and there they were abundant. We could not even find a dried carapace of an Apus, in fact the only ones secured during the Expedition were two dilapidated specimens found by Mr. Watt and myself during our return journey along the Stevenson Creek, but as the termination of the abdomen was wanting it was impossible to say whether they had belonged to the genus Apus or Lepidurus.

We were very disappointed at not securing the Estheria-like animal alive but collected a number of the carapaces though these alone were not sufficient for purposes of identification. Fortunately as previously said I secured a few specimens of the entire animal during my subsequent visit to Charlotte Waters just after the rains had fallen, a year later, and then also obtained another closely allied species of which not even the carapaces were to be seen at Coulin Lagoon. Though Mr. Squire has carefully searched for the animal in the same spot during the two recent seasons he has not been able to find a single living specimen. The animal belongs to a new genus closely allied to Estheria and Limnadia and has been described by Mr. Hall and myself under the name of *Limnadopsis squirei*, the other species secured along the Stevenson being called *L. tatei*. The genus is not however confined to the central region as Professor Tate had previously collected a few carapaces of another species (*L. brunneus*) in the Northern Territory.

The periodicity of occurrence of certain animals in this central area of the continent has already been alluded to and is well shown in the case of Limnadopsis. Amongst the Crustacea there are certain species which always seem to be obtainable after rain and certain others which are not so certain to appear in any particular spot, though they may previously have been collected there in large numbers.

Take the clay-pans about Alice Springs for example during the past three seasons. Apus appears to have always been present at the right time. Once, three years ago, *Limnadopsis squirei* was abundant and has not been seen since though carefully searched for; *Estheria packardi* on the other hand is always present and persists in its three varieties, var. *typica*, *cancellata* and *minor*, as long as muddy pools remain. *Limnadopsis tatei* has not yet been found

At Charlotte Waters and in the neighbourhood Apus is always to be found for a short time; *Estheria packardi* in abundance and *Estheria lutraria* may be relied upon.

Last year (1895) *Limnadopsis squirei* and *L. tatei* were found but have not apparently put in their appearance this year (1896), whilst a recent gathering made by Mr. Byrne contains a new species of Limnadia which was certainly not to be found in the clay-pools there during the previous year. Of course the forms not met with in the pools searched may be and probably are developed elsewhere, but it shows how certain forms are dominant and seems to suggest a greater power of adaptability on their part to such influences perhaps as variation in length of drought and it is at the same time worth noting that the constantly recurring, dominant forms, *e.g.*, Apus and Estheria spp., are just those which have bright red blood, whilst the forms of irregular occurrence, Euliminadia and Limnadopsis spp., are strongly contrasted with the former when the two series are swimming about together by the absence of red blood and their general pale colour.

The staff at Alice Springs was considerably interested in the "barking" spider as it was called, though the word booming better expresses the nature of the sound which it was supposed to make. The spider (*Phlogius crassipes*) was found without any difficulty by the blacks close to the station, where in hard sandy ground it makes its burrows. Each of these is about an inch in diameter and goes down in a slanting direction for about two feet, when it terminates in a little more or less spherical chamber in which are the remains of beetles and a small amount of webbing and in which the animal remains during the day time. There is no protective covering for the hole on the surface.

In addition to listening at night close by the burrows in which we knew the spider was living, and to keeping it alive in captivity in variously shaped receptacles some of which were made so as to resemble as nearly as possible the shape of the burrow with its swollen termination, Mr. Besley, a member of the station staff, and myself spent a night out in the bush in a spot where it was plentiful, hoping to settle the question.

We heard the noise attributed to the spider, but came to the conclusion that it was made by a bird—probably a quail. It is a noteworthy fact that the noise is principally heard at the time when birds, such as quails, are most abundant. I could find no structure which could enable it to make any such noise as is attributed to it, but at the same time our observations of the animal in captivity led to the discovery that it does possess a well developed stridulating organ. When irritated it rises on its hind legs, and rubbing its palps against its maxillæ produces a low whistling sound. The structure of this organ is described and figured in the Zoological section of the Report. There is a series of little stiff rods on the maxilla which rub across a series of curious little flattened "keys" on the palp and so produce the low whistle. The animal is closely allied to a spider from Assam, in which Professor Wood Mason many years ago described a very similar organ, and since then, in fact since we found the organ in *Phlogius crassipes*, it has been shown by Messrs. Pocock and F. O. P. Cambridge that stridulating organs of various kinds are more common than was previously thought.

All the specimens captured while I was staying at Alice Springs and since then have, unfortunately, been females, so that we do not know if the organ exists in the male, but from the fact of its occurrence in the former it is perhaps to be regarded as an organ for producing a warning signal to warn off would-be aggressors. At the same time it must be said that this is merely a theory which has not been put to the test, as we do not know either who its particular enemies are or whether they are capable of hearing the sound made. If they be, as probably they are, ground animals such as the smaller marsupials or lizards it may, especially in the case of soft-bodied animals through whose skin its powerful jaws can bite, act as a deterrent. In its burrow we found remnants of beetles upon which it had evidently been feeding. During the night, as it is nocturnal, it is doubtless active, but during the day time when taken out of its burrow it is very sluggish, and can easily be handled.

What with the days spent either in examining and sketching the animals brought in by the blacks, or in collecting out amongst the hills and the evenings in developing photographs taken during the day and in long talks and discussions on anthropological subjects with Mr. Gillen, some of whose valuable notes relating to the customs of the Arunta tribe are published in the Report of the Expedition, my time at Alice Springs soon passed away. Mr. Watt had returned from his flying visit to the gold and "ruby" fields, and at midnight on 5th August, with the temperature below freezing point, we left the station on our southward journey along by the overland telegraph line.

It will be many years before the recollection of our stay at Alice Springs fades from our memory, for it came as a pleasant ending to an Expedition which had carried us into parts of the continent remote from the usual beaten tracks.

Looking back upon our Expedition a few scenes stand out prominently—the gibber plains at sunset ; the bare upland stony plain with the thin telegraph line streaking away to the horizon, on which through the heated air waves the outline of the Charlotte Waters Station can be seen ; the view of the great Finke Valley where at Crown Point the river breaks through the Desert Sandstone hills ; Chambers Pillar rising solitary amongst the sandhills ; the picturesque water-holes of the George Gill Range ; the camp, weird and silent, by Lake Amadeus ; Ayers Rock glowing bright red in the sunset ; the group of graceful palm trees by the side of the rock-pools in Palm Creek and the wonderful gorges amongst the McDonnell Range.

Six days' incessant travelling—camping out in the open wherever we happened to come to some time after dark and starting away at sunrise—brought us to Charlotte Waters. After spending a few hours here with Mr. Byrne we started off again, and in three days more reached the head of the railway line at Oodnadatta and three days later we were in Adelaide.

SUMMARY

OF THE

Zoological, Botanical and Geological Results of the Expedition.

By BALDWIN SPENCER, M.A., C.M.Z.S., Professor of Biology in the University of Melbourne.

SUMMARY

OF THE

ZOOLOGICAL, BOTANICAL, AND GEOLOGICAL RESULTS OF THE EXPEDITION.

By BALDWIN SPENCER, M.A., C.M.Z.S., Professor of Biology in the University of Melbourne.

CONTENTS.

	PAGE
Zoology	139
Botany	159
Geology and Palæontology	162
General Conclusions	171

At the suggestion of Professor Tate, to whose work I am already much indebted, I have added to the narrative the following short summary of the results of the Expedition so far as they are concerned with Zoology, Botany and Geology. I have not included in the summary the Anthropological work for the simple reason that a mere brief outline of the work of Dr. Stirling and Mr. Gillen would have been of no value, whereas, in the case of the three sciences mentioned the connection between them is so intimate and the bearing, especially of the Geological work, upon the important question of the distribution of the fauna and flora is of such a nature that it appeared to be of advantage to bring together and briefly discuss the main results arrived at in these departments. For the Geological and Botanical results I am of course indebted to the papers written by Professor Tate and Mr. Watt, singly and in conjunction with each other. On one or two points, such as for example the previous existence of a "cosmopolitan flora" in Australia, I have ventured, when discussing certain general conclusions, to differ from the views put forth in the reports.

ZOOLOGY.

In the narrative some of the more interesting points in regard to different forms of animals found have been already alluded to. I shall here endeavour to briefly summarise the general results.

The following table indicates the number of genera and species of animals the occurrence of which in the Centre is recorded in the Zoological section*:

	No. of Genera.	No. of Species	No. of New Species
Mammalia.			
Carnivora	1	1	0
Chiroptera	2	2	0
Rodentia	3	11	3
Marsupialia	16	24	6
Monotremata	1	1	0
Aves	83	100	5
Reptilia.			
Lacertilia	22	11	10
Ophidia	11	13	2
Amphibia	1	6	1
Pisces	5	8	5
Mollusca	20	38	21
Arthropoda.			
Crustacea	7	11	5
Lepidoptera	11	22	1
Coleoptera	125	177	66
Araneidæ	35	57	18
Orthoptera	33	59	19
Hymenoptera	18	31	8
Vermes.			
Oligochaeta	1	1	1
Total	398	603	171

The above list does not include certain forms collected but of which the examination has not been yet completed. Amongst these may be noted the Hemiptera (of which a considerable number were collected), Myriapoda, Scorpionidæ, Pseudo-scorpionidæ (two species), Diptera, Isopoda (one species), Turbellaria (one species of water planarian), Hirudinea (one species), Rotifera (one species of Lacinularia) and Polyzoa (one species).

* In the table are included certain Marsupialia, Lacertilia and Hymenoptera, recorded in the Supplement to the Zoology, published in Part I.

In certain groups, especially in the Coleoptera, numerous species have been already described, but excluding the latter the list given indicates within narrow limits our present knowledge of the numerical proportions of the Central Australian fauna.

That the numbers will be steadily increased in time is of course certain, as many of the rarer forms can only be secured at intervals and under very favourable circumstances such as the successive occurrence of two or more good seasons.

It is highly probable, in fact certain, that the fauna varies to a great extent with the climate. Central Australia may be described as possessing a permanent and a fluctuating fauna ; the former, which may be regarded as the nucleus of its fauna, consists of species which have become especially adapted to life in an arid region ; the latter consists of immigrant species not so hardy and only to be met with when more favourable seasons have rendered their immigration from out lying regions possible.

Probably the permanent fauna is fairly well represented in the collection made and amongst certain groups, such especially as the Land Mollusca, there is no fluctuating fauna, but in the case of others, such as the Insecta especially and to a lesser extent the Mammalia, the fluctuating fauna, dependent as it is primarily upon the vegetation, is an important factor. With a succession of bad seasons the vegetation dwindles and the animals, except the most hardy species, disappear, and even the latter become very much thinned out. With a recurrence of good seasons first of all the surviving inhabitants increase in numbers, and then if the good seasons last long enough a gradual immigration takes place.

The permanent fauna again may be divided into two groups, the first containing those animals which can always be found during the dry season, the second containing those which only appear during the short wet season.

The collector who sets to work as we did in a dry season, especially if he has been accustomed to the moister coastal district, is first of all struck with the fact that there is a wonderful poverty of animal life except so far as regards ants, flies, grasshoppers and certain beetles, birds and lizards. He naturally misses almost all forms of life associated, as the Phalangeridæ for example, with well wooded districts, or the Platypus with the sheltered pools of permanent rivers, and he rapidly appreciates the influence of a climatic barrier.

After turning over every available stick and hundreds of stones and finding no trace of moisture he realizes how impossible it is for creatures such as land

Planarians or Peripatus or even the wide-spread land Amphipod to exist in such a region.

Next he becomes wearied with the unsuccessful search after insects on flowering shrubs, though such as Cassias and Eremophilas are abundant and attractive enough, and day after day he finds the same forms of life. Every water-hole, every loamy plain or sandhill yields a wearying, monotonous and small series of animals until he begins to realise that the fauna is characterized by the entire absence of the rich series of species of the coastal districts and the presence of a relatively few dominant species which are evidently capable of adapting themselves to conditions of the most unfavourable description for animal life.

In the wet season the fauna changes as if by magic, insects formerly unseen come about in swarms, fresh water Crustacea crowd the clay-pans and water-holes, caterpillars in thousands creep about, the majority of them simply falling a prey to the lizards, frogs and birds which increase with like rapidity. At the same time though animal life is now abundant it is composed of relatively few species, each existing in enormous numbers. Probably towards the close of the favourable season a horde of migratory rats will pass like a wave across the country, disappearing into the depths of the desert, where they perish. For a time the small marsupials will be more or less abundant, but soon they also will disappear to æstivate during the dry season or only to come out from their hiding places during the cool of the night—the majority of them probably perishing before they reach maturity.

Rapidly the country assumes its dry state, and the only animals left are the hardier forms which can withstand the heat and dryness, and the few inhabitants of the deeper and scattered water-holes.

If the drought be abnormally prolonged then even the hardiest animals will suffer, and the fauna will be so reduced that it may take some time before increased fertility on the part of the survivors and the influx of immigrants from the broad belt of land enclosing the central region will make good the deficiency.

Probably, what is certainly true of the plants, holds good in the case of animals, and that is that the struggle for existence is not of such a complicated nature as in many other parts. After the rain falls, a caterpillar or insect or frog for example has no lack of food—there is plenty for all—though of course they are each liable to fall a prey to birds or reptiles or mammals. There does not seem indeed to be any attempt made except perhaps on a most limited scale, at anything like protective colouration. Grasshoppers and insects crawl about in thousands without any attempt at concealment—every animal of every kind seems so to speak to forget all else except the necessity of feeding as rapidly as possible and reproducing its species.

The first phase in the struggle for existence is concerned with the development of the ovum. Unless the development be very rapid the animal has no chance of growing to the size at which it can take advantage of the rapidly disappearing food supply—disappearing not because there is not enough and to spare for all but because the vegetation on which all depends can only withstand the temperature for a given length of time. The second phase in the struggle is entered upon when the dry season supervenes, and this is really dependent upon the first, for it is only those who have grown to a certain size and who in addition have hardy enough constitutions who have any chance of lasting out the drought with its miserably small supply of food and water.

One important point in connection at all events with the smaller marsupials and probably with all the animals to a greater or less extent is, as already noticed in the Zoological report,* "that they attain full size at very varying periods of life and that an animal reared during a successive series of bad seasons and consequent dearth of food may never attain the full size characteristic of the species, though at the same time it may bear young ones." This curious fact is well seen in the case of *Phascologale cristicauda* where the smallest mature male measures (head and body) 136 mm. and the largest 220 mm. ; the smallest female of the same species measuring 125 mm. and the largest 170 mm., though both of the latter were carrying young ones in the pouch. In each case the larger forms were obtained at the close of a good season. In the same species the number of teats varies between four and eight, the latter being present again in those captured at the close of a good season.

These facts will serve to show the direct influence which the climate has upon the development of the animals, for no such relatively great variations exist amongst allied species found in the coastal districts where the climate is not liable to such irregular fluctuations.

Taking the different groups we may now point out the more important points concerned with each.

Mammalia.—In the Eutheria the most important forms are the large bat (*Megaderma gigas*) and the Rodents. The former is only found in the caves

---

* Part II., p. 25.

amongst the ranges and in Central Queensland. Amongst the Rodents eleven species so far are determined, including the introduced *Mus musculus*. Only one species *Conilurus* (*Hapalotis*) *mitchelli* is known from West Australia also. Two species of Mus (*M. gouldi* and *M. greyi*) are widely distributed save on the west, but the characteristic Rodents of the Centre are those belonging to the genus Conilurus (Hapalotis) which includes the jerboa-rats. One of these (*C. pedunculatus*) is known as yet only from the Higher Steppes, the other four are characteristic of the eastern and central parts of the interior.

These rodents have a remarkable habit of travelling periodically in vast hordes. In the middle of 1895, for example, Mr. Byrne, writing from Charlotte Waters, said, "The jerboa-like rodents are coming from the eastwards and they almost amount to a plague here." Two months later scarcely one was to be seen. This migration *from the East* appears to show that the Centre receives periodic additions to its fauna dependent primarily upon the seasons, and that in the case of the rodents, as their distribution indicates, the immigration takes place from the East.

The most interesting form—evidently rare, as only a single immature specimen was secured amongst the large number of rodents caught—is a species of Mastacomys. The interest of this lies in the fact that the genus is represented by a single species living in Tasmania and by a fossil form from the Wellington Caves in New South Wales. In the Centre it has only been found at Alice Springs amongst the ranges. Evidently it represents an old form of Rodent and is one of the very few animals which Tasmania and the Centre have in common. In this respect it stands in strong contrast to the characteristic rodents of the genus Conilurus, which is not represented in Tasmania and may, like the Diprotodonts, be regarded as having originated on the eastern side of the continent, whence they have spread out westwards.

The marsupials are represented by twenty-three species, which may be divided into three groups :—

1. A few species widely distributed over the continent. These include *Trichosurus vulpecula*, *Sminthopsis murina*, *S. crassicaudata*, *Perameles obesula*.

2. A larger number which are characteristic of the inland parts of the eastern divisions (Queensland, New South Wales and Victoria) and of South and West Australia. These include *Macropus robustus*, *Macropus rufus*, *Petrogale lateralis*, *Onychogale lunata*, *Lagorchestes conspicilatus* var. *leichardtii*, *Bettongia lesueuri*,

Chœropus castanotis, Phascologale calura, Dasyurus geoffroyi, Phascologale cristicauda, Antechinomys laniger, Myrmecobius fasciatus.

3. Those which as far as yet known are peculiar to the central region. These include *Peragale minor,** *Perameles eremiana, Sminthopsis psammophilus, S. larapinta, Phascologale macdonnellensis, Dasyuroides byrnei, Notoryctes typhlops.*

The species included under the second head may again be divided into two series: (*a*) those widely distributed over the interior from east to west (*M. rufus, M. robustus, P. lagotis* (?), *O. lunata, C. castanotis, Phascologale cristicauda* (?), *P. calura, Antechinomys laniger* (?), *D. geoffroyi*); (*b*) those restricted to the west side (*Petrogale lateralis, L. conspicillatus, Bettongia lesueuri*).

In addition to these positive features there are equally striking negative features in the absence of many representative genera of other parts—an absence often, but by no means always, associated with the lack of arboreal vegetation. Such genera are, for example, Dendrolagus, Æpyprymnus, Potorous, Dromicia, Petaurus, Pseudochirus, Phascolarctos, Phascolomys. Perhaps so far as specific affinity is concerned, in the case of the marsupials just as in that of the rodents, the most striking fact is the presence of only one species common to the Centre and Tasmania and that is the ubiquitous *Perameles obesula.*

The most distinctive marsupial of the Central region is without doubt *Notoryctes typhlops* and it must be confessed that the modification of this curious creature to adapt it to a burrowing life in hot, sandy country—if this modification be regarded as having taken place within the time during which the Centre has undergone desiccation—is a most remarkable one, all the more so because it is a modification without parallel in any other marsupial of the district. It seems indeed preferable to suppose that Notoryctes is the modified survivor of some perhaps extinct burrowing marsupial similar in its habits to the true mole. On the other hand it is quite likely that there may yet be found in some of the large, incompletely explored parts of the continent one or more allied existing forms whose burrowing habits have hitherto caused them to escape detection.

Next to Notoryctes the most characteristic marsupials are the species of the genus Peragale; all three of these, viz., *P. lagotis, P. leucura, P. minor,* are found in the Centre to which probably the latter two are confined, while the first is also a Western and South Australian form.

---

* Probably also *Peragale leucura.*

If we divide Australia into two parts, one including the coastal districts on the north, east and south-east and the other including the rest of the continent, then we can regard the Marsupial fauna of the Centre as an assemblage of species belonging to those characteristic of the second region which have become adapted to life in the more arid parts.

Of the two families of the Monotremata only one—the Echidnidæ—is represented. The species is the common continental one, *Echidna aculeata* var. *typica*, the range of which, as in the Centre it is found from Burrow Springs in the north to Charlotte Waters in the south, is now shown to extend over the whole continent.

Aves.—The birds represent 100 species, of which five are new. In addition to these the most interesting and important specimens are those of *Spathopterus alexandræ*, the Princess of Wales Parrakeet. This is closely allied to the genus Polytelis, in which the species had previously been placed. Mr. North has placed it in a separate genus characterized by the fact that in the adult male the third primary feather is much elongated and terminates in a spatule.

With regard to the distribution of the species obtained Mr. North says: "The majority of the birds collected range over the southern half of the Australian continent from east to west, but there is a slight preponderance of western forms. Several north-western species are now recorded for the first time from Central Australia; but it is worthy of note that no strictly northern species is represented in the collection."

Lacertilia.—The collection of Lacertilia is perhaps the most representative of the series as it was supplemented by important additions secured during the wet season, when not only are some of the rarer forms more plentiful, but others not seen at all during the dry season can be obtained. As Messrs. Lucas and Frost point out* it contains four groups:—

1. A series of widespread species. These include *Delma fraseri*, *Lialis burtoni*, *Amphibolurus barbatus*, *Varanus gouldii*, *Egernia whitii*, *Hinulia lesueurii*.

2. A series of western species. These include *Amphibolurus maculatus*, *A. imbricatus*, *A. reticulatus*, *Moloch horridus*, *Tympanocryptis cephalus*, *Egernia stokesi*, *Rhodona gerrardii*, *R. bipes*, *Ablepharus greyi*; while *Amphibolurus pictus*, *Tiliqua*

* Zoological Report, Part II., p. 112.

occipitalis and *Tympanocryptis lineata* extend across the south of the continent from West Australia to the interior of Victoria.

3. A series of northern forms. These include *Heteronota bynoei, Nephrurus asper, N. laevis, Diplodactylus ciliaris, Physignathus longirostris, Varanus giganteus, V. punctatus, V. acanthurus, Hinulia fasciolata*.

4. A series peculiar so far as yet known to the Central district. These include *Ebenavia horni, Ceramodactylus damaeus, Diplodactylus byrnei, Tympanocryptis tetraporophora, Diporophora winneckei, Varanus eremius, V. gilleni, Rhodona tetradactyla, Ophidiocephalus taeniatus*.

The affinities with the western species are the most marked. Out of thirty-eight species no fewer than twenty-two are found in Western Australia.

The next greatest amount of affinity lies with the Northern Territory and North Queensland, which have between them fourteen in common with the Centre. Victoria and New South Wales only share with the Centre some five ubiquitous species, while with Tasmania there are not more than two species in common.

In the case of the Lacertilia we see again, as in the Marsupials, a marked line of distinction between the interior and the south-eastern coastal fauna, a still more strongly marked affinity between the Centre and the west, and unlike the Marsupials, a strong affinity with the north.

One result of the large series of forms secured has been the discovery of a very large amount of variation in forms closely allied but hitherto considered to be distinct from one another, and in consequence of this the merging of certain species. Thus, for example, *Nephrurus lœvis* and *N. platyurus* are merged in the former species; the opinion of Dr. Gunther that *Heteronota derbiana* and *H. bynoei* are variations of the one species is confirmed; *Tympanocryptis tetraporophora* serves to connect *T. lineata* and *T. cephalus*, while the large series of *H. lesueurii* cause Messrs. Lucas and Frost to say,* "Thus we shall be prepared to include as varieties of the last named *H. spaldingi*, Macleay (= *H. dorsalis*, Blgr.), *H. leœ*, Blgr., *H. strauchii*, Blgr., *H. inornata*, Gray, *H. essingtonii*, Gray, *H. muelleri*, Fischer, and *H. tœniolata*, White."

Amongst the vertebrata the lizards, as might have been expected from the nature of the country, form the most striking part of the fauna, and probably there yet remain a considerable number of species to be obtained, but these lie amongst the rarer and less widely dispersed forms.

* *Loc. cit.*, p. 113

Whilst some are widely distributed over the whole region amongst the loamy and sandy flats of the Lower Steppes and on the broad valley and even hillsides of the Higher Steppes others are more or less characteristic of one or other of these districts, and others again are still more local and appear to live in small colonies occupying a very restricted area.

The most widely diffused forms are *Gehyra variegata* which is always to be met with under logs and the bark of trees, *Amphibolurus reticulatus* burrowing everywhere on sandy and loamy ground and perhaps the most abundant of all forms; *Amphibolurus barbatus*, even more widely spread than the former but not nearly so numerous; *Moloch horridus* and the ubiquitous *Egernia whitii* and *Hinulia lesueurii*, while *Egernia stokesi* is found on the hard loamy plains of the Higher and Lower Steppes but not in soft sandy country or upon the ranges.

Certain species, on the other hand, though they extend to a certain extent on to the Lower Steppes are characteristic of the Higher Steppes; such for example are *Nephrurus asper*, *Varanus giganteus*—the largest of Australian lizards, which lives in caves and holes amongst the higher ranges—*Varanus punctatus* and *V. acanthurus* and amongst the smaller forms the four species of Ablepharus.

The Lower Steppes are characterised by the following series, some of which again spread to a certain extent on to the Higher Steppes. Amongst the common forms are *Amphibolurus pictus*, which though it extends farther north is peculiarly characteristic of the southern part, and *Tiliqua occipitalis*.

Amphibolurus maculatus, the most brilliant in colouration of all the lizards, is very restricted in distribution, occurring in one or two colonies along the Finke and the same applies to *Varanus eremius*, a ground form.

Rhyncœdura ornata is known in the Centre only from the neighbourhood of Charlotte Waters, from which also come the two most interesting species secured during the Expedition, viz., *Ebenavia horni* and *Ceramodactylus damœus*.* The former is the representative of a genus containing only one other species in Madagascar—a distribution which calls to mind that of the genus Casuarina amongst plants.

Ebenavia is distinguished from other genera of the Geckonidæ such as Diplodactylus by the absence of claws and is most closely allied to Phyllodactylus, a genus not represented in the Centre but with three species in West Australia,

---

* *Ceramodactylus damœus* has since been recorded by Messrs. Lucas and Frost as occurring in Northern Queensland.

one of which (*P. marmoratus*) extends into South Australia and the interior of Victoria. Probably the distribution of Ebenavia will be found to extend into West Australia.

Equally curious is the distribution of Ceramodactylus, which is only recorded hitherto from Persia and Arabia. In the Centre it is found only along the Finke near Charlotte Waters.

The Ophidia are too imperfectly known to make it safe to draw any conclusion from the small series obtained. It is quite possible that the number of species in the Centre is very few, but future work will probably considerably increase the number yet known. Out of the twelve species secured one (*Hornea pulchella*), the representative of a new genus, is known only from Charlotte Waters; *Furina ramsayi* is the commonest form in the southern part extending across to West Australia and to the interior of New South Wales, whilst a new species, *Haplocephalus stirlingi*, is widely distributed from Alice Springs in the north to Oodnadatta in the south.

Amphibia.—The Amphibia are remarkable, as might perhaps have been expected in such a region, by (1) the paucity of species and (2) the great numbers in which at certain seasons the individuals of the species represented are found. At the present time some sixty-five species are known in Australia. Only six are recorded from Central Australia.

In contrast to the Marsupials and Lacertilia there is very little affinity between the Amphibian fauna of West and Central Australia; the only common species is *Hyla rubella*, which is also found in New South Wales, Queensland and the Northern Territory.

The characteristic Amphibia of the Central region consist of five species (*Limnodynastes ornatus, Chiroleptes platycephalus, C. brevipalmatus, Heleioporus pictus, Hyla rubella*), four of which may be described as burrowing frogs and they are also inhabitants of the interior of the east and south-east parts of the continent. They may either represent the direct descendants of forms which inhabited the region during most favourable climatic conditions—a supposition which is probably true in regard to *Hyla rubella*—or they may be species which have immigrated from outlying eastern and south eastern parts in comparatively recent times, which is probably true in the case of *L. ornatus, Chiroleptes platycephalus, C. palmatus* and *Heleioporus pictus*, which are elsewhere burrowing frogs and so are capable of migrating across country dry for the greater part of the time. The last-named species extends from the Victorian coast (Melbourne) through South Australia and into the Centre.

Out of the six species three are characteristic of the Lower Steppes—*Chiroleptes platycephalus, C. brevipalmatus, Heleioporus pictus*—and were not met with amongst the ranges of the Higher Steppes. They seem to prefer the hard, sandy and loam plains where the water-holes are periodically dried up and where it is absolutely essential for them to burrow, their power of storing water in their bodies being of considerable service to them. *Limnodynastes ornatus* follows the sandy river beds where it can burrow down to moisture throughout the Lower and the Higher Steppes as well. *Hyla gilleni* is restricted to the north and is of very rare occurrence and probably an immigrant from the north, while the home of the little *Hyla rubella* is undoubtedly amongst the water-pools in the ranges, whence in flood time it is periodically washed down to supply the water-holes on the Lower Steppes, the wells sunk at intervals along the overland track enabling it to survive where otherwise it would perish.

Pisces.—Pisces are represented in the collection by eight species belonging to six genera. Out of the eight species six are new, viz., *Nematocentris tatei, Nematocentris winneckei, Eleotris larapintæ, Gobius cremius, Chatoessus horni, Plotosus argenteus*; of the remaining two *Therapon truttaceus*, Macleay, is known from the Endeavour River, and *Therapon percoides*, Gunth, from Queensland rivers. One of the most striking features amongst the fish is the absence of affinity with those of the Murray River system. The genera Oligorus, Ctenolates, Murrayia, Gadopsis and Copidoglanis of the Murray are entirely wanting. The genera represented are those of coastal districts and may perhaps be best regarded as having been derived from the north.

To the eight species must be added another, viz., *Therapon fasciatus*, recorded from "near the McDonnells" by Mr. A. H. S. Lucas, which is of interest as being a West Australian species.

Mollusca.—The Mollusca is in many respects the most interesting of the orders represented as it contains considerably the greatest proportion of endemic species. The number of land molluscs known to inhabit the region has been increased from three previously recorded to twenty-five, of which, according to Professor Tate,* four only extend beyond the area and five are close allies of

---

Zoological Report, Part II., p. 182.

species found outside the area. The fresh water molluscan fauna has been increased from one to thirteen species.

Speaking of the land mollusca Professor Tate says, "The facies of the fauna approximates more to that of sub-tropical and temperate West Australia than of any other part of the continent, and is in strong contrast with the highly differentiated fauna of tropical and sub-tropical Queensland, situated to the east of the Cordilleras, to which it is geographically equally near. The limited number of genera represented, together with the facts of their geographic distribution, would seem to indicate a primitive population, which has been maintained in an isolated condition by climatic and geologic changes."

Of the species described *Thersites fodinalis* is recorded from West Australia and the interior of New South Wales, *T. perinflata* from West and South Australia, *Microphyura hemiclausa* from North Queensland and the Northern Territory, *Pupa contraria* from West Australia, *Succinea interioris* from the interior of Queensland.

Taken as a whole the distribution accords well with that of other groups such as the Marsupials; the main affinity lies with West Australia, whilst *T. fodinalis*, *T. perinflata*, and *S. interioris*, show affinity with the interior part of the continent lying to the east. There is the same striking contrast between the molluscan fauna of the Centre and that of Northern and Tropical Queensland on the one hand and the south-eastern coastal districts, including Tasmania, on the other, which in reality is the leading feature of the whole central fauna.

At the present time there can be no passage of land Molluscs either into or out of the Central region, which has been in this respect isolated since Pliocene times. To account for the relationships of the land Mollusca we must postulate what there is abundant evidence of—a more favourable climate in Pliocene and earlier times allowing of migration from the west, north and to a lesser extent the east, both into and out of the Central area.

Whilst the Rolling Downs formation (Upper Cretaceous) was being deposited the Central highlands formed a large island mass. Professor Tate says: "At this period a more or less cosmopolitan fauna and flora prevailed, and it was doubtlessly then that the Larapintine area acquired its species of Microphyura, Charopa and Flammulina and those species of a more or less maritime habitat belonging to Liparus, Stenogyra, Pupa and Succinea. How else is it possible to account for the presence of about eight species of land snails in the very centre of the continent absolutely isolated from allied or identical species which are to-day cir cumferential in their occurrences? The insularity of its geographic position was partially maintained during the deposition of the Desert Sandstone (Upper Cretaceous*)—a fresh water area, or largely so, replacing the maritime one. Favourable conditions then ensued in Pliocene times which permitted migration over the largely reclaimed lacustrine areas. It was then that *Badistes perinflata* and *B. jodinalis* spread south and south-west; so also the Angasellae but under new modifications; whilst there may have been received a few northern types, out of which have been evolved *Badistes granditubereulata, B. wattii, Chloritis squamulosa, Thersites subleerata* and *T. adcockiana*. The final climatic phase was the creation of the Dry Zone, which effectually cuts off migration in a southerly direction."

It may perhaps be pointed out that these suggestions with regard to the times at which the Molluscan fauna was established in the Centre are not altogether satisfactory. Thus Professor Tate in the paragraph previous to the one quoted says "the Endodontae and Flammulina belong to genera largely Tasmanian." His suggestion that the species of Microphyura, Charopa (Endodonta) and Flammulina were acquired when the Larapintine Region formed an insular mass is somewhat difficult to understand, as if the Centre were in this insular state then it could have but little chance of giving or receiving Mollusca to or from any other part and especially in the south-easterly direction, such as any Tasmanian affinity would imply.

It seems scarcely necessary to go back so far to find a time at which the special forms mentioned by Professor Tate passed across into or out of the Centre. There can be no doubt, as Professor Tate and others have repeatedly pointed out, that in or about Pliocene time the climatic conditions of the Centre were favourable to animal life. If as Professor Tate suggests it was then that *Badistes perinflata* and *B. jodinalis* spread south and south-west and that "there may have been received a few northern types, out of which have been evolved *Badistes grandi tubereulata*, etc.," why will not the same favourable time suffice for the migration of such forms as *Microphyura hemiclausa*, which is specifically identical with the Queensland and Northern Territory form?

What it would appear, judging not only from the Molluscan but from the Marsupial and Lacertilian fauna, to be necessary to postulate, is that the centre has been connected, in such a way that emigration of animal life was fairly easy across the intervening country, with (1) the north and north-east, and (2) with the

---

* This is referred to as Supra Cretaceous in the Geology Report and the Rolling Downs as Upper Cretaceous.

west, and that further the latter connection has been more marked than the former, and so has exerted a stronger influence. This might be brought about by the western connection persisting for a somewhat longer time than that on the north and east; possibly the western connection was established before the eastern, and existed also during the time of the latter.

Mr. Hedley who, in the Appendix to the Mollusca Report, has dealt with the anatomical features of a certain number, has kindly allowed me to reproduce from his correspondence with me on the subject the following interesting and suggestive extracts. Mr. Hedley says: "The Mollusca point clearly to an original population derived from Western Australia, composed of Xanthomelon,* Liparus, Pupa, Succinea. Then an immigration primarily from the northern territory, but remotely from Queensland, is shown by Thersites, Microphyura, Stenogyra, Bithinia, Melania and Corbicula. This migration from Queensland probably occurred when the Queensland fauna was far poorer than it is now, and as Microphyura and Thersites travelled from Queensland, the west sent in exchange the few Xanthomelon which have reached the Pacific coast, and which are still confined to the north. After this, communication with Queensland ceased, so that the rich fauna which lately poured across Torres Straits from New Guinea failed to reach even in one instance to Port Darwin.

"Another striking lesson to be learnt from an analysis of the fauna is the impenetrable barrier which shut out Tasmanian types. Not a single member of that numerous, active, most enduring group the Rhytididae has reached this region. Originating in Antarctica, one colony occupied New Zealand and spread thence through New Caledonia to the Solomons. Another established itself in Tasmania and marched in force to Cape York and even crossed to Mount Owen Stanley in New Guinea. Yet the enormous lapse of time and change of land and water requisite for these wanderings was not sufficient to allow Rhytididae to pass into the Larapintine Region."

Amongst the land molluscs a very clear distinction occurs between the Lower and the Higher Steppes; out of twenty-five found in the latter only three extend southwards into the former; these are (1) *Thersites perinflata* a widely distributed species ranging over the interior from the Burt Plain in the north to the Flinders Range in the south, and westwards to the Fraser Range and Yilgarn in Western Australia, and (2) *Pupa contraria* which just passes into the northern part of the

---

* It will be seen from the Appendix to the Mollusca Report that the generic name of Xanthomelon is applied by Mr. Hedley on anatomical grounds to certain species classified by Professor Tate, respectively, in the genera Angasella, Chloritis and Thersites.

Lower Steppes and is also recorded (its original locality) from the Houtmans Abrolhos off the coast of Western Australia, and (3) *Thersites jodinalis* the most abundant land shell.

Whilst such species as *Microphyura hemiclausa*, *Angasella setigera*, *Thersites adcockiana*, *Stenogyra interioris*, *Pupa moorcana* and *Succinea interioris* are distributed largely amongst the ranges of the Higher Steppes, others such as *Endodonta æmula*, *E. planorbulina*, *Flammulina retinodes*, *Angasella euzyga*, *A. winneckeana*, *A. arcigerens*, *Chloritis squamulosa*, *Thersites granditubcrculata*, *Thersites wattii*, *Liparus spenceri* and *Pupa ficulnea* are very sporadic in occurrence, most of them being so far as yet known limited to one single locality and often occurring in small colonies in a space not more than a few yards square.

As Professor Tate says, "Like the truly endemic plants, the land snails live on the southern escarpments of the elevated land or in the deeply-shadowed gorges of the same and occur in very restricted areas, sometimes as one colony only, or if in more, then usually widely separated from one another."

With regard to the water mollusca, *Melania venustula*, *Bithinia australis*, *Bulinus pectorosus*, *Corbicula sublævigata*, and *Unio stuarti* are only recorded from the Lower Steppes; *Melania balonnensis*, *Bulinus texturatus* and *Isidorella newcombi* occur in both the Lower and the Higher Steppes, while *Limnæa vinosa*, *Bulinus dispar*, *Planorbis fragilis*, *Ancylus australicus* and *Sphærium translucidum* are recorded only from the Higher Steppes.

Crustacea.—In the Crustacea the Phyllopoda are represented by nine species belonging to five genera, the Macroura and the Brachyura by one species each.

Up till the time of the Expedition no species of Apus was definitely recorded from Australia though its existence has been mentioned by Messrs. Sanger and Banckroft, but as no description was given it was impossible to determine whether this species belonged to the genus Apus or its close ally Lepidurus, which has long been known from the continent and from Tasmania and New Zealand.

The species *Apus australiensis* was first secured near Charlotte Waters in the Centre, and its distribution is now known to extend over the interior of the colonies of New South Wales and Queensland and into West Australia. Just as Apus is characteristic of the interior and west, so Lepidurus, which is not found there, is characteristic of the eastern coast and of Tasmania and New Zealand. The New Zealand, New South Wales, Victorian, Tasmanian and (with perhaps some doubt in the case of *Lepidurus angasi*) the South Australian Lepidurus are referable to one species, *L. kingii*.*

Amongst the Limnadiadæ the genus Estheria is represented by three species of which one, *E. packardi*, with its three varieties *typica, cancellata* and *minor*, is by far the most abundant and is widely spread over the Lower and the Higher Steppes, occurring in every water-hole except the cold, deep and clear pools amongst the ranges. The various species of Limnadiadæ all seem to require muddy water for their existence and long after the others have died out (except perhaps an occasional *E. lutraria*) and are represented by empty carapaces, *E. packardi* in one or other of its varieties will be found surviving. *Estheria lutraria*, described originally by Brady from a single dried carapace secured by Professor Tate from Cooper's Creek, is confined to the water-holes of the Lower Steppes, whilst *Estheria dictyon* was only secured along the Palm Creek in the James Range.†

Limnadopsis is a new genus "distinguished from Estheria by the presence of a haft-organ; from Limnadia and Eulimnadia by the spinous processes on the dorsal edge of the carapace, by the different number of lines of growth and of pairs of feet; from Limnadella by the difference in size of the two pairs of antennæ."‡ The larger species of this genus (*L. squirei*) measures fully three-quarters of an inch in length and is probably the largest of the living Estherianæ.

Amongst the Crustacea the genus forms the most distinctive type in the Central region, though it is also represented in the Northern Territory by a species (*L. brunneus*) of which carapaces were collected by Professor Tate near Port Darwin.

In the Macroura the single species *Astacopsis bicarinatus* is widely distributed over Australia. It apparently owes its wide distribution to its capacity for burrowing.

In the Brachyura the occurrence of *Telphusa transversa* is a striking feature. It only occurs in the water-holes amongst the Lower Steppes and was not met with in the Higher Steppes or anywhere along the Finke River. There can be no doubt as to its identity with the form described by Professor Haswell from Thursday Island. In all likelihood it will yet be found in the interior of

---
* This statement is based upon a careful examination by Mr. Hall and myself of a large series of forms from various localities.
† This species occurs also in the collection of the South Australian Museum, labelled only "S. Australia."
‡ Geological Report, Part III., p. 239.

Queensland, and it may perhaps be best regarded as one of those forms such as the burrowing frogs which have been able to make their way into the Centre owing to their capability of burrowing and so of surviving during periods of drought.

Lepidoptera.—The collection of Lepidoptera was of necessity a small one, as it was made during the winter months when only a few were to be obtained. Save during the early part of the Expedition no insects were even attracted to the light at night time and the collection of Lepidoptera except in the case of a very few forms was practically impossible. The most plentiful forms which were widely spread over the district were the introduced *Danais petilia* and the ubiquitous *Pyrameis cardui* (var. *kershawi*).

Coleoptera.—In the Coleoptera (exclusive of the Carabidae) the same difficulty was experienced as in the case of Lepidoptera—the time of year was unfavourable. The eight hundred specimens secured represent one hundred and forty-five species, of which sixty-two are new, and of these four are referred by Mr. Blackburn to new genera. In the Carabidae thirty-two species were collected, of which four are new.

A considerable number of species of Coleoptera have already been recorded from Central Australia, and I am indebted to the Rev. T. Blackburn for the following general note upon the Coleopteran fauna of Central Australia:—

"It is very doubtful whether the facts hitherto ascertained in respect of the *Coleoptera* of Central Australia are sufficiently numerous to warrant any general conclusions founded upon them. Most persons (myself included) who have visited what is commonly called the "stony desert" to investigate the subject have found the *Coleoptera* very rare but have been informed by residents that at some season of the year (other than the then present season) they are very plentiful. A request, however, to procure and forward a large batch, at the time of plenty, leads to small results (in several instances I am satisfied that this has not arisen from unwillingness to take the requisite trouble). But the conclusion it would be natural to draw from such experiences is probably incorrect, for on the occasion most favorable to the resolution of the matter (viz., the residence for about six months at one locality in Central Australia of Mr. A. Zietz, a scientific collector, though not especially a Coleopterist, nor able to devote much time to the Coleoptera) the local tradition was verified by the observation of Mr. Zietz that on certain occasions, usually I understand the brewing of a thunderstorm, Coleoptera were abundant, although at ordinary times there were very few to be met with. Mr. Zietz sent to Adelaide very large numbers of specimens, and, I think, more numerous species than have been taken by any other collector in Central Australia.

Probably, whenever certain conditions are fulfilled, specimens usually in hiding come forth and are seen in plenty, and probably the conditions necessary are conditions unfavourable to casual collectors being on the alert, so that many such persons have been once or twice accidentally in contact with such plenty, but might never be again.

Having thus qualified the value of opinions relating to the Coleoptera of Central Australia, I may, with less fear of misleading, venture to report on some of the general characteristics of that group of insects so far as they are at present known. The prevalent type is certainly, I think, South Australian, but with a tendency to extreme specialisation, and with a certain admixture of tropical forms. As might be expected from the scarcity of vegetation, ground beetles are much more numerous than Phytophagous species, but in group after group of both classes the species are very distinct from those found in other parts of Australia, not so much in general appearance as by structural peculiarity and the number of isolated genera is comparatively large. Considering the dryness of the country it is surprising to find that the *Hydrophilidæ* are comparatively numerous. *Carabidæ*, *Tenebrionidæ* (especially *Helœides*) and *Curculionidæ* (especially *Amycterides*) are the prominent groups of the Coleopterous fauna; and there are a somewhat large number, usually highly specialised, of *Lamellicornes*. Large size and bright colouring are rare among the Coleoptera of Central Australia. I have observed a prevalence distinctly greater than in other Australian fauna to extremely *pallid* colouring. The few *Buprestides*, even, that have been found in Central Australia are (with scarcely an exception) among the less attractively coloured species of their genera."

Araneidæ.—In the Araneidæ the 150 specimens collected are referable to fifty-seven species belonging to thirty-six genera. Eighteen species are described as new, one being the representative of a new genus. Out of the series at present known from Central Australia, thirty-one are recorded from the eastern colonies, the great majority of which are known from Queensland and New South Wales, two (*Epeira extuberata* and *Migas paradox* are known from New Zealand only, one (*Idioctis helva*) is recorded only from Fiji, one only (*Habronestes scintillans*) is common with West Australia, whilst *Latrodectes scelio* extends from the centre through Queensland, New South Wales and Victoria, and to the north Island of New Zealand. It is not at present possible to draw any conclusions with regard to the distribution of spiders in Australia as the great majority of those hitherto described have been collected in Queensland and New South Wales, the western and southern part of the continent having been but little explored so far as this group is concerned.

The most interesting forms are the new species, *Nephila eremiana*, the large orb webs of which, extending across as much as ten or twelve feet from tree to tree, form a prominent feature in the more open scrub, and the Queensland *Phlogius crassipes*, the largest of the Australian tunnel-forming spiders, which is interesting as possessing a well-developed stridulating organ.

Hymenoptera.—In the Hymenoptera, of which only thirty-one species are described, the most interesting forms are (1) the little black, yellow-footed ant forming its nest out of sand particles fastened together with the resinous secretion of the porcupine grass, and (2) three species of "honey-ants." The porcupine-grass ant has been described by Mr. W. F. Kirby as a new species under the name of *Hypoclinea flavipes*, and judging by the way in which the curious and characteristic so-called "galleries" which are always made by it in the country traversed by us, have been described from various parts of the interior, the species is probably widely spread over and at the same time peculiar to the interior of the continent. Its distribution may very likely be coterminous with that of the resin producing species of Triodia.

In the "honey ants" Mr. Froggatt describes three species, two of which are new. *Camponotus inflatus*, Lubbock, is evidently widely spread over Central and West Australia, whilst as yet *C. cowlei*, Froggatt, and *C. midas*, Froggatt, are only known from restricted areas amongst the central ranges.

Oligochæta.—Of earthworms only a single species is known, which is referable to the genus Acanthodrilus. As already described,* the sporadic distribution of the species in the centre together with the absence of genera at present characteristic of the more coastal parts of the continent point to the fact that the genus Acanthodrilus was more widely spread over the north-eastern part of the continent in former times, when there was no such climatic barrier as now exists separating the centre from outlying parts, than it is at the present day. In all probability the species of Microphyura and Acanthodrilus passed across to the Centre from the north at the same time, and with the change in climate which succeeded Pliocene times have been isolated.

* Zoological Reports, Part II., p. 116.

BOTANY.

The list of plants recorded by Professor Tate numbers 611. Prior to the Expedition the number described from the region was 502. The additions consist of 8 new species, 16 species new for South Australia and 112 species new to the region.

In his paper* "On the Influence of Physiographic Changes in the Distribution of Life in Australia" Professor Tate suggested, mainly on botanical grounds, the division of Australia into three regions—(1) Autochthonian, (2) Euronotian and (3) Eremian. The latter region occupies the central area of the continent and is coterminous with that over which the annual rainfall is under 10 inches; westward it extends to the coast line of mid West Australia. For the central region occupied by the table land of Ordovician sandstone from which rise the long parallel ridges forming the McDonnells and the James Range the name Larapintine is used adapting the native name of the Finke River the basin of which lies mainly within the area. To the south lies what Professor Tate calls the Central Eremian district, stretching south from about the latitude of Engoordina (Horseshoe Bend) on the Finke and formed by the Cretaceous table-land.

The latter area is practically the same as that referred to in the narrative as the *Lower Steppes*, the Larapintine region being comprised within the *Higher Steppes*.

In regard to the Larapintine Flora Professor Tate concludes that after the deposition of the Rolling Downs Formation, which isolated except perhaps in a northern direction the Larapintine table-land (or the Higher Steppes), a lacustrine area was formed during the period of deposition of the Desert Sandstone, and that a cosmopolitan flora prevailed at this period which continued into Paleocene times. Pluvial conditions continued into Pliocene times, whilst in Post Pliocene times a high state of desiccation was reached. Under these changed conditions the original "cosmopolitan" flora became largely extinct and an immigration of Oriental forms supervened. The present flora of the Eremian region has been "developed from Autochthonian and Euronotian elements and largely modified by Oriental immigrants and the species evolved from them."

The Larapintine flora is described by Professor Tate as follows:—

I.—Exotic Species, chiefly Oriental	125
II.—Endemic species of Exotic genera	219
III.—Endemic species of Australian genera	270

* Aust. Ass. Adv. Sci., vol. i., pp. 312-325, 1889.

The first two groups, together with 52 in the third, comprise plants belonging on the whole to the Eremian type, while the remaining 218 species of the third group "are either actually Autochthonian or Euronotian or are related species, and as a whole may be viewed either as residues of a common Australian flora or as modified descendants therefrom."

Professor Tate accepts the views of Baron von Ettingshausen with regard to a cosmopolitan flora " which originated in Late Cretaceous times in Europe, North America and Australia."

In his recent presidential address to the Linnean Society of New South Wales Mr. Deane has drawn attention to the grave doubts which exist as to the validity of the conclusion drawn by Baron von Ettingshausen and others, and it is more than probable that this supposed "cosmopolitan flora" with fossil remains of genera such as Quercus, Alnus, Betula, Salix, etc., in Australia will have to be abandoned.*

The plants of the Larapintine region, so far as their habitat is concerned, are divided by Professor Tate into two groups: (1) the *Lowland* vegetation and (2) the *Saxatile* vegetation.

The Lowland vegetation comprises that of the river banks, loamy plains and sandy ground. Its species are widely diffused through the Eremian region, spreading far south in South Australia, over the internal parts of New South Wales and Southern Queensland and westwards to the coast line of mid West Australia. Thus the Lowland vegetation of the Centre has no less than sixty per cent. of its species common to the flora of both Shark's Bay and Nichol Bay on the West Australian Coast. Its species "are either immigrants from the Oriental Botanical province or are endemic species of extra-Australasian genera."

In both the Lowland and Saxatile plants the truly Australian forms are as a general rule characterized by their sporadic distribution. They are "frequently

---

* Proc. Linn. Soc. N.S.W., vol. x., 1896, p. 216. At the close of a valuable summary of work dealing with this question, Mr. Deane says (p. 655): "At present the facts seem to afford grounds for concluding —

1. That many, if not all, the typical Australian floral types originated in Australia or in some land connected with it, but now submerged.

2. That the assumption of the existence of a universal flora of mixed types at any epoch is unfounded.

3. That the fossil plant remains of Tertiary age in Eastern Australia indicate a vegetation in all respects similar to that existing on the coast in the same latitude at the present day.

To these might perhaps be added a fourth conclusion of less certain character, but of high probability, that the *Proteaceæ* represent a most ancient type which had their origin at a time when not only extensive areas of land existed in the Southern Hemisphere but when some kind of connection more or less lasting existed between Australia and South Africa."

gregarious in isolated colonies, sometimes occupying a few square yards, or even as much as several square miles." The alien plants on the other hand are widely distributed and able to adapt themselves to extremes of soil and climate.

The saxatile vegetation growing on the ranges in crevices on the escarpments and especially on the rocky sides of the deep and shaded gorges supplies the greater number of the characteristic Larapintine species. Thus "of seventy flowering plants, restrictedly rock dwellers, seven only are of exotic origin."

A striking feature of many of the species is either their isolation or sporadic distribution. The fan-palm, for example (*Livistona Mariæ*), is limited to a single colony along the Finke gorge and a small tributary, the Palm Creek: *Swainsonia canescens* was only seen growing in two small colonies a few yards square and separated from one another by nearly eighty miles, the grass tree (*Xanthorrhœa Thorntoni*) occupies a narrow belt of country seventy miles long by thirty wide, and other species in the same way occurred only in single or in very few colonies, often far apart.

In the case of certain species we have as it were connecting links between the Autochthonian and Euronotian floras. *Hibbertia glaberrima* for example is the only species of the section Hemihibbertia extending beyond Western Australia and its distribution in the centre connects it with the same species in Queensland; *Gastrolobium grandiflorum* in the same way stretches across from the west to the interior of Queensland and New South Wales, and the same applies to other species such as *Styphelia Mitchellii* a species found in Queensland whilst the home of the genus is in West Australia.

In the *Central Eremian region* the prevelance of Salsolaceous plants is a striking feature, their place being taken in the Larapintine area by grasses, the most important of which are species of Triodia or "porcupine" grass which extends also over large areas of the true desert region stretching across to West Australia. Other characteristic plants of the Central Eremian area are *Cassia eremophila* and *Eucalyptus microtheca* which in the Larapintine district are replaced by *C. phyllodinea* and *E. rostrata*.

Atriplex rhagodioides, Salsola Kali, Kochia aphylla, Bassia diacantha and species of Acacia such as *A. aneura* (Mulga), *A. cyperophylla* (red Mulga) and *A. homalophylla* (Gidlea) are common and on loamy patches such plants as *Lepidium papillosum, Euphorbia Drummondii*, and species of Ptilotus.

Speaking generally we may regard the present flora of the centre of the continent as consisting of two distinct elements (1) a series of forms which are the descendants of those which occupied the area when under more favourable climatic conditions than now exist, it was possible for plants to spread across both from the west (Autochthonian), and from the east—especially the north-east (Euronotian). This flora derived partly from the west and partly from the east spread across as the land gradually rose after the deposition of the Upper Cretaceous beds, and whilst, over a wide region of the centre, the Desert Sandstone formation was being deposited in lacustrine areas. Later on, in Pliocene times, the hygrometric conditions still allowed of an inter-communication between the east and west across the centre, but in Post-Pliocene times, with the gradual desiccation, the original flora was slowly extinguished, its representatives lingering only in favoured spots. (2) The second element consists of a series of more hardy species from the oriental region, which gradually spread southwards, until finally the remnants of the original flora survived only in the shady gorges and escarpments of the mountain ranges.

GEOLOGY.

In the reports by Messrs. Tate and Watt the various geological formations of the area traversed are described under the following heads Pre-Cambrian, Ordovician, Post-Ordovician Conglomerate, Upper Cretaceous, Desert Sandstone (Supra-Cretaceous), Tertiary.

(1) *Pre-Cambrian.*

These comprise the series classed as Pre-Silurian by Mr. Chewings and as Archean by Mr. H. Y. L. Brown. Travelling northwards along the overland track "a sudden and striking change is observable in the lithological character of the rocks at the point where those of Pre-Cambrian age succeed the Lower Silurian, four or five miles south of Alice Springs Telegraph Station. Leaving quartzites and limestones we find ourselves among rocks of a highly metamorphic character, such as gneisses and schists of various kinds." To the north these rocks extend to the Burt Plains forming an irregular series of rough, broken hills. East and west, where the junction line between the Pre-Cambrian and Lower Silurian rocks can be seen, the latter, resting unconformably on the former, form a prominent ridge with a steep northern escarpment. In the McDonnell Ranges alone the rocks now described by Messrs. Tate and Watt as Pre-Cambrian are estimated to occupy an area of at least 10,000 square miles, and the "region clearly furnishes an almost typical example of regional metamorphism in which great changes, both physical and chemical, have been produced in the rocks by earth-movements." The evidence obtained points to the eruptive origin of a large part of the metamorphic group whereas the Cambrian rocks of Australia, as far as at present known, are entirely sedimentary.

In regard to the distinct stratification and definite and determinable dip of the rocks described by Messrs. Brown and Chewings the conclusion is arrived at that "although it may be possible and even in places probable that the planes, which are so strongly developed, coincide with the original planes of stratification in any large area where sedimentary rocks may have been developed, yet as a general rule there can be no doubt that these planes represent foliation planes. This statement is greatly strengthened by the facts of the coincidence over large areas of the strike of these planes, and of their great persistency; for they are traceable not only through rock-masses, the eruptive origin of which is highly probable, but also even through undoubted intrusive dykes. They are therefore planes of foliation, of stratification-foliation—that is of foliation corresponding with the original bedding planes, it may be in places, but elsewhere assuredly they appear to be those of cleavage foliation."*

In regard to the age of the rocks† it is pointed out that the strong unconformity separating them from the Lower Silurian group shows them to be either Pre-Cambrian or Cambrian. In lithological character and tectonic structure they differ from the known Cambrian strata of Yorke's Peninsula and Flinders Range, and agree apparently with the Pre-Cambrian rocks of the Mount Lofty Range. Whilst no eruptive dykes have been noted in Central Australia amongst the Lower Silurian rocks, they are very numerous amongst the Pre-Cambrian and exhibit different stages of metamorphism in the same district, which tends to show that they have been intruded at different periods. If the highly metamorphosed rocks were of Cambrian age then some of the eruptive dykes which appeared last might have been expected to have penetrated the Ordovician strata. Lastly, whilst the Cambrian rocks of Yorke's Peninsula, the Flinders Range and the Kimberley district are fossiliferous there is an entire absence of fossils in the metamorphic rocks of the centre.

(2) *Ordovician.*

To this horizon Messrs. Tate and Watt refer the strata forming (with the exception of the Post-Ordovician conglomerate to be mentioned later) the series of ridges which run roughly parallel to one another from east to west across the

---

* Part III., Geology, p. 40.   † *Loc. cit.*, p. 37.

centre to the south of the Pre-Cambrian area. "Beginning from the north these comprise the quartzite ridge which forms the southern boundary of the Pre-Cambrian area and in which are the Heavitree, Emily, Temple Bar, etc., gaps. This ridge is succeeded on the south by the Waterhouse, James, George Gill, Levi and Chandler Ranges. They have a mean combined width, if we include the intervening plains and valleys of from sixty to seventy miles. The area occupied by them, therefore, must be more than 15,000 square miles."*

Certain of these strata have been previously assigned by Messrs. H. Y. L. Brown and Chewings to the Cambrian period but the subsequent discovery of fossils of Ordovician age in certain of these rocks and of waterworn fragments of Ordovician limestone containing characteristic fossils in others show that this determination was erroneous.

In 1891 Professor Tate referred certain fossils obtained by Mr. Chewings at the head of the Walker River, Merrenie Bluff and Petermann Creek to the *Upper Silurian*, but in the same year Mr. R. Etheridge, Junr., referred fossils secured by Mr. H. Y. L. Brown from the same horizon to the *Lower Silurian* age, and the latter author then referred the rocks of the George Gill, the James and the Ooraminna Ranges to the same age.

This determination of the Lower Silurian age of the fossil bearing rocks has been confirmed and adopted in the report.

Messrs. Tate and Watt now eliminate Cambrian from the classification of the rocks and "include in the Ordovician system all the strata lying between Mount Burrell cattle station on the south and the McDonnell Ranges on the north, with the exception of the conglomerate which was observed on the north side of Rudall Creek and on the banks of Ellery Creek north of the Lutheran Mission Station (Hermannsburg)."

The Ordovician rocks consist for the most part of quartzites and sandstones with beds of limestone, clay-slate, micaceous slates and sandstone. Thus for example in the section (Geology, Plate I., Fig. 5) across the McDonnell Range in the neighbourhood of Mount Sonder and south to the Missionary Plains, the Pre-Cambrian gneiss and mica-schist are seen lying to the north of the range. The high ridge is formed mainly of Ordovician quartzite replaced on the south by micaceous clay-slate, underlying which are thick beds of magnesian limestone which pass to the south under the river alluvium, forming the valley along which flows the Davenport Creek. Gneissic granite outcrops in this valley representing

* Part III. Physical Geography, page 5.

probably an inlier of Pre-Cambrian rocks. To the south of this valley rise two parallel ridges of quartzite, enclosing between them the Horn Valley, along which outcrops a band of limestone the existence of which has probably determined the line of denudation which has given rise to the valley. Forming the northern boundary of the broad Missionary Plain and resting unconformably upon the quartzite of the southern of the two ridges just mentioned lies a bed of Post Ordovician conglomerate.

The Ordovician strata have been thrown into a series of folds, those of the northern part having been subjected to greater disturbance than those of the south.

Thus the Levi Range consists of sandstones dipping at low angles—about 8 or 10 — to the south on the north side and at about the same angle to the north on the south side, the range being thus formed out of a gentle synclinal trough. In the north, in the James Range and at Mount Sonder for example, the strata have suffered much greater disturbance, the quartzites dipping at very high angles.

The folding has been produced along lines running in a general east and west direction and "the chief factors in addition to the position of the longitudinal valleys occupying the original troughs of the folds, that have influenced the direction of the lines of denudation are (1) the lines of weakness on the crowns of the anticlinal arches and (2) the position of the bands of limestone. An example of the influence of (1) is furnished by the valley of the Petermann Creek, which has been eroded out of an anticlinal arch, while the rocks of the corresponding synclinal trough now form the George Gill and Levi Ranges. The influence of (2) as might have been expected is to be observed throughout this region, the greater number of the valleys within these ranges having been, to a great extent, eroded out of the limestone beds."*

The gorges and gaps through which the main stream flows across the successive ridges, with rocks of quartzite and sandstone rising almost vertically to heights varying from 200 to 800 feet above the valleys, owe their origin to the fact that the erosion of the river beds in the position of the present gaps kept pace with the upheaval and folding of the strata. By a gradual lowering of their channels as the rocks rose the streams have been able to maintain their original course, so that the characteristic feature of the streams flowing over the Ordovician area is the fact that they do not follow the trend of the main valleys but run at right angles to these.

* Part III., Physical Geography, p. 6.

(3) *Post-Ordovician Conglomerate.*

This conglomerate flanks the southern face of the quartzite ridge which forms the northern boundary of the Missionary Plains and the southern boundary of the Horn Valley. The lower parts of the conglomerate consist of fragments derived from the Ordovician strata and in this pebbles of red limestone were obtained containing the following Ordovician fossils:—*Actinoceras tatei, Palæarca wattii, Orthis dichotomalis*. The Post-Ordovician age of this conglomerate was thus clearly established. The upper layers were most largely made up of pebbles derived from the Pre-Cambrian rocks, and the total thickness of the conglomerate and conglomeratic sandstone was estimated by Mr. Watt, who carefully examined it during a traverse of the ranges, to be not less than 7000 feet.

(4) *Upper Cretaceous.*

The Cretaceous plains and table-land slope gradually from their northern limit somewhere in the neighbourhood of Mount Burrell Station, where they have an elevation of not less than 1000 feet to Lake Eyre in the south where their surface is thirty-nine feet below sea level.

From these stony and loamy plains rise table topped hills capped with Desert Sandstone.

The table-land formation is recognised as contemporaneous with the Rolling Downs series of Queensland, which has been assigned by Messrs. Etheridge and Jack[*] to the Lower Cretaceous series but, according to Messrs. Tate and Watt, "the facies of the fauna is more akin to that of the European Upper Cretaceous while the palæontological differences between it and the Desert Sandstone are too slight to justify the application of the terms Lower and Upper to them respectively." The Rolling Downs formation and its equivalent series forming the table-lands and plain of the Central area are therefore recognized as *Upper Cretaceous*. The formation is essentially an argillaceous one and the Oodnadatta bore which reaches a depth of 1571 feet shows a series "varying from clay shale to marly clay intercalated with which are thin argillaceous limestones and some sand beds; those latter occur at various horizons, and the chief supply of water was obtained in the basal sands of the section. Thus the general character of the strata passed through is like that of other bore sections in the Lake Eyre basin."[†] Above the level plains rise low hills of which Mount Daniel with an elevation

[*] Geology of Queensland, etc., p. 380.   [†] Geology, Part III., p. 62.

of 1330 feet above sea level may be taken as an example. Beneath the Desert Sandstone capping (eighteen feet) are purple and grey shale twenty-two feet; red shale, forty feet and beneath them an unknown thickness of yellow and grey shale.

Passing northwards towards the James Range, that is towards the old shore line of the Cretaceous sea, the shales and clays are replaced as might have been expected by sandstone. "For the most part the stratification of the Upper Cretaceous is apparently horizontal, though slight undulations of far reaching extension prevail in the northern area occupied by the rocks."*

With regard to the supply of Artesian water in the Cretaceous area which has been dealt with in important papers by Messrs. Etheridge,† Jack‡ and Brown§ the conclusion is reached owing to the "far northerly extension of the Cretaceous rocks and the replacement of the prevailing argillaceous condition by sandy strata towards the northern boundary" that it is probable that in the district traversed " the source is, after all, of local origin." The Finke in its course from Henbury to Crown Point and the Goyder and Lilla Creeks near their sources flow approximately along the line of junction of the sandy Cretaceous and the impermeable Ordovician limestones. In this way, especially as the Cretaceous beds have a slight southerly dip the flood waters may be absorbed and carried down to considerable depths in the depressed Lake Eyre basin and so provide the supply obtained by such bores as those at Oodnadatta, Hergott and Strangways.

(4) *Desert Sandstone* or *Supra-Cretaceous*.

The greatest thickness of this formation as seen at Crown Point was estimated at fifty feet. It consists there of "sharp grains of glassy quartz, varying much in size, cemented by opaque-white highly siliceous matter and more or less stained red by oxide of iron."‖ The identity of the formation over wide areas of the interior from South Australia to Queensland has previously been clearly pointed out by Messrs. Jack and Etheridge¶ and Mr. H. Y. L. Brown.** By Messrs. Jack and Etheridge the Rolling Downs are regarded as Lower Cretaceous, the Desert Sand stones as Upper Cretaceous. Mr. Brown on the other hand describing the Cretaceous strata between the 139th parallel and the western boundary line of Queensland from Lat. 26 to Lat. 32 S., says, as quoted by Messrs. Jack and

---

Geology, Part III., p. 61.
† ‡ Geology of Queensland, etc., pp. 414-433
§ Aust. Ass. Adv. Sci., Brisbane, vol. VI., 1895, p. 330
§ Aust. Assoc. Adv. Sci., Sydney, vol. i , 1878, p. 243.
‖ Geology, Part III., p. 65
* Geology of Queensland, etc. The Desert Sandstone Formation, p. 511.
  Report of Gov. Geologist Adelaide, 1883, etc.

Etheridge, "the strata consist of brittle clays and calcareous shales with bands of limestone and gypsum, clay, ironstone, and ferruginous sandstone and sandy beds . . . *overlying this formation* are beds of sandstone, argillaceous sandstone, kaolin, grit and pebbly conglomerate forming table lands and hills almost invariably capped by a thin bed of yellow and red flinty quartzite or jasper rock the total thickness varying from one hundred to two hundred feet . . . The composition of these Super-Cretaceous beds is the same over wide areas from the Warrego in New South Wales, to the Diamintina." Messrs. Jack and Etheridge say: "It will be seen that Mr. Brown does not distinctly aver that the "Super-Cretaceous" rocks described by him lie unconformably on the Cretaceous ; there can be no doubt however that he so understands their relations, as is evident from the section accompanying the report. The identity of the "Super-Cretaceous" of South Australia with the Desert Sandstone of Queensland in Mr. Brown's mind is settled by his remark that the Grey Ranges of New South Wales and Queensland belong to the same formation. The "porcellanised" condition of a portion of the sandstone on the South Australian side of the border is a very interesting observation in view of the "quasi-vitreous" appearance of the formation at Cloncurry and Croydon on the Queensland side.

The superposition of Tertiary Rocks on the Desert Sandstone of South Australia is an observation of the highest importance, as direct evidence of this nature is quite wanting in Queensland, and Daintree ascribed a Tertiary age to the Desert Sandstone itself."*

Messrs. Tate and Watt agree with Mr. Brown in assigning a Supra Cretaceous age to the Desert Sandstone.

Whilst no fossils have actually been recognised in the Desert Sandstone of the Finke Basin, plant impressions have been reported by Mr. Brown, Professor Tate and others as occurring, together with marine molluscs, in the Desert Sandstone of the basins of Lakes Eyre, Frome, Torrens and Gairdner. In addition to the only two plant remains previously assigned to the Desert Sandstone,† viz., *Didymosurus* (?) *gleichenioides* and *Glossopteris* sp., Professor Tate now adds ten more and states that "the flora here indicated is analagous with that at Vegetable Creek and Dalton, described by Baron von Ettingshausen, and on palaeontologic ground has been regarded by him as Eocene. The same type of flora is preserved at various localities in Victoria, the age of which is considered by McCoy to be

---

* Geology of Queensland, etc., p. 550.   † Geology of Queensland, p. 531.

Miocene."* Messrs. Hall and Pritchard† however have shown that certain plant bearing beds in Victoria lie below the marine Eocene, "and this," says Professor Tate, "accords well with the general fact that wherever the base of the marine Eocene is reached lacustrine and plant-bearing beds succeed in depth."

In the section dealing with Post-Cretaceous Phenomena the question of the silicification of the Upper and Supra-Cretaceous rock is dealt with. Messrs. Jack and Etheridge,‡ in referring to the deposition of the Rolling Downs and Desert Sandstone, point out that the latter must at one time have occupied at least three-quarters of the present surface of Queensland, though now its denuded remnants only cover about one-twentieth of their original area. After the Rolling Downs formation had been laid down a considerable upheaval took place. "The denudation of the Rolling Downs formation followed and must have gone on for some time. Unequal movements of depression then brought about lacustrine conditions on portions of the now uplifted bottom of the old sea strait, and in other portions permitted of the admission of the waters of the ocean. Finally a general upheaval placed the deposits of the period just concluded in nearly the positions in which we now find them."

Messrs. Tate and Watt point out that after the deposition, first of the Upper Cretaceous, and then of the Supra-Cretaceous (Desert Sandstone), both series underwent a considerable amount of denudation *before* the silicification, which is now so characteristic a feature of the latter, took place. "In every example of silicification of the sediments of Upper Cretaceous age there is no covering bed, and when the Desert Sandstone is present the alteration is limited to that formation. It may therefore be inferred that denudation of the Cretaceous plateau preceded the process of silicification, which acting from above downwards affected whatever sediment chanced to be at the surface." The greatest amount of silicification is seen between the Stevenson River and Charlotte Waters in which district also the the largest number of obsidian bombs and unrolled agates are found.

The origin of the silicification is very difficult to account for. At present two theories have been advanced (1) Mr. East§ has supposed that it is due to deposition from silicated waters, the siliceous material being derived from the decomposition of the metamorphic rocks of the McDonnell Ranges and that the silicification took

---
* General Geology, p. 67, 68.
† Aust. Ass. Adv. Sci., vol. v., Adelaide, 1893, p. 358.
‡ Geology of Queensland, etc., p. 511.
§ J. J. East "On the Geological Structure and Physical Features of Central Australia." Tran. R.S. S. Aust., vol. xii., pp. 31-53. 1889.

place during the later stages of depositions of the Desert Sandstone whilst Messrs. Tate and Watt have argued that a considerable interval must have elapsed between the formation of the Desert Sandstone and the silicification ; (2) Messrs. Tate and Watt point out that the formation of agates and obsidian bombs and of the Desert Sandstone Breccia require a common origin and suggest that though there are great difficulties in the way of its acceptance because of the "widespread silicification and the actual absence over its area of any traces of actual volcanic outbursts" it is essential to assume the former existence of volcanic action, the silicates of the ash beds or larva being the source of the requisite siliceous material. "The obsidian bombs demand volcanic action . . The development of agates within the volcanic material was only another phase of siliceous precipitation. Of this supposititious volcanic formation all that remains are the agates and the obsidian bombs."

Tertiary.—Professor Tate has on previous occasions drawn attention to the fact that in what he terms the newer Pliocene times pluvial conditions prevailed over the central area. Indications of this are to be seen in the form of gravels through which the present river channels have cut their way, and in the form of terraces along the margin of the broad valleys along which now wander the reduced water courses. At that time Lake Eyre must have been an inland sea, and fossil remains prove that it was inhabited by alligators now extinct (*Pallimnarchus pollens*), and by such fish as Ceratodus, while the land was inhabited by a marsupial fauna consisting of genera such as Diprotodon, the larger number of which are now extinct. The river Goyder close to where we crossed it ran between cliffs about thirty feet high composed of river detritus ; on the north side of the escarpment at Crown Point, a well-defined shingle beach rises to an elevation of fifty feet, while three miles south of this the Yellow Cliff, fifty feet high, bounding the southern bank of the water-course, consists of tumultuously bedded sandstones and conglomerate.

The former existence of a considerable lacustrine area is shown also by a fossil deposit at Dalhousie, which has the nature of a gypsiferous tuff containing numerous shells of *Melania venustula*, *M. lutosa*, *M. balonnensis*, *Bithinia australis* and *Corbicula sublævigata*. None of these were found in the waters of the mound spring close to which is the deposit, nor are they found living in the immediate neighbourhood, while *M. lutosa* does not now occur in the central area.

Palæontology.—In regard to this the more important facts as detailed by Professor Tate are as follows.

Fossiliferous limestone beds, of the same horizon, were met with at Hpilla gorge near Tempe Downs, near Petermann Creek and close to Laurie Creek. Away to the north the same series outcrops along the whole length of the Horn Valley. In conjunction with and underlying these beds, are fossiliferous quartzites and sandstones; in addition the quartzite of Chandler Range and Mount Watt yielded fossils.

The uppermost zone is rich in *Orthis læviensis* while below this are beds rich in Trilobite remains, the most important palaeontological find of the Expedition being that of an entire *Asaphus illarensis*, Eth. fils., as up to the present time though Trilobites have been described from the beds only fragments have been found. In addition to this species three others were secured, viz., *Asaphus thorntoni*, Eth. fils., *Asaphus howchini*, Eth. fils., and *Asaphus lissopeltis*, Tate, the last-named being a new species.

In addition to the Trilobites the limestones yielded numerous species of Orthoceras and Endoceras whilst the limestones, sandstones and quartzites yielded a remarkable preponderance of Isoarcae, imparting to this fauna a local feature.

Not a trace of graptolites was discovered. So far as the correlation of the fossil bearing rocks is concerned Professor Tate is of opinion, though the proofs are not conclusive, that " there is presumptive evidence that the Gordon River group (*i.e.*, of Tasmania) and the Larapintine series are contemporaneous and younger than the Victorian graptolite slates," and he is also inclined on account of its representative fauna to regard the Larapintine series as the equivalents of the Caradoc series of England.

GENERAL CONCLUSIONS.

The origin and relations of the present flora and fauna of Central Australia are intimately bound up with the past history of the continent, firstly as regards its relationship to other land masses and secondly as regards the changes which have taken place in the form of the continent and the relations of the various parts of the present land area. In his presidential address to the Australasian Association for the Advancement of Science held in Adelaide in 1893, Professor Tate has given a valuable summary of our geological knowledge of the interior of the continent. He has shown that Sturt was the first to surmise the fact that in Pliocene times there existed pluvial conditions when the southern part of the central area centering in the Lake Eyre district was occupied by a great inland sea or series of fresh water lakes. The existence of these rendered possible, though not known to Sturt, the large series of now extinct marsupials such as Diprotodon, and it also made possible a connection between the western and the eastern parts of the continent.

For a fuller account of the various workers who have dealt with this subject the reader is referred to the above-mentioned address of Professor Tate, whose own work is undoubtedly the most valuable which has been published during the past few years in connection with the past history of Central Australia. For many years also Professor Tate* has been engaged in collecting and collating information with regard especially to botanical feature, and the result of these he embodied in his presidential address to the Biological Section of the Australasian Association in Sydney in 1888 entitled "On the Influence of Physiographic Changes in the Distribution of Life in Australia." In that address, after reviewing the geological changes which are now known to have taken place in Central Australia and their influence on the distribution especially of plants within the limits of the continent, Professor Tate, mainly on botanical grounds, proposed the division of the Endemic Australian flora into three types and of the continent into three corresponding regions:—

1. *Euronotian*, occupying the coastal area on the north, east and south-east, its internal boundary coinciding with the rainfall limit of 25-50 inches per annum.

2. *Autochthonian*, a small region restricted to the south-west corner of the continent, its internal boundary also coinciding with the same rainfall limit in this part.

3. *Eremian*, occupying a large stretch of country, centering in Lake Eyre but extending right across the continent to the shores of Western Australia, and over which the average rainfall is less than ten inches per annum.

In regard to the Euronotian region Professor Tate says that the type flora of this is "dominant in the south and east part of the continent."

Mr. Hedley, at the Australasian Association meeting held in Adelaide in 1893,† in his paper entitled "The Faunal Regions of Australia," pointed out that the regions suggested as suitable in the case of plants were not equally satisfactory when applied to animals. Accepting the Autochthonian and Eremian regions he suggested the division of the Euronotian into two, for one of which, including

---

* Aust. Ass. Adv. Sci., vol. i., Sydney, 1888, p. 312.
† Aust. Ass. Adv. Sci., Adelaide, 1893, vol. v., p. 444.

Tasmania, Victoria and southern New South Wales the name Euronotian should be retained, whilst for the second, including Queensland and northern New South Wales, he suggested the name Papuan.

There is no doubt but that Mr. Hedley's division of the Euronotian into these two parts is essential so far as zoology is concerned, in fact such a division was already hinted at by Professor Tate in respect of botanical features in the remark that the Euronotian type was dominant in the south and east part of the wider region to which he applied the term.

In dealing with the question of the relations of the flora and fauna of the various parts of Australia as we find them existing at the present time, perhaps the point of most importance is the demonstration of the fact that for a long period of time the east and west parts of the continent were separated from one another by an impenetrable barrier of some description. Mr. Hedley says :* " Owing to fundamental errors of his interpretation of Australian geology, Wallace's treatment of the subject in ' Island Life' is of but slight value." It is quite true that, owing to the imperfection of our geological knowledge when Mr. Wallace wrote, he was mistaken in suggesting that a great inland Tertiary sea acted as a barrier, but whilst this is so, the main facts of central importance were most clearly enunciated by Wallace who, arguing from a knowledge of Sir Joseph Hooker's work, wrote :† " These facts again clearly point to the conclusion that south western Australia is the remnant of the more extensive and more isolated portion of the continent in which the peculiar flora was principally developed. . . But whilst this rich and peculiar flora was in process of formation, the eastern portion of the continent must either have been widely separated from the western or had perhaps not yet risen from the ocean. During some portion of the Secondary period therefore this (i.e., the east) side of Australia must have been almost wholly submerged beneath the ocean ; and if we suppose that during this time the western part of the continent was at nearly its maximum extent and elevation, we shall have a sufficient explanation of the great difference between the flora of Western and Eastern Australia, since the latter would only have been able to receive immigrants from the former, at a later period, and in a more or less fragmentary manner."

Whilst the more exact nature of the barrier and of the successive geological changes occurring in the central area of the continent since Cretaceous times have been demonstrated by other workers, notably by Messrs. Etheridge and Jack,

---

*loc. cit., p. 444.   † " Island Life, 1st edit. 1880, p. 464.

Professor Tate and Mr. H. Y. L. Brown, there can be little doubt but that the work of Mr. Wallace in regard to the distribution of animals and plants in Australia is second in importance to that of no other writer, both in relation to its suggestiveness and to the conclusions which he draws, though at the same time, with increasing knowledge, it may be necessary to modify certain of the latter.

What seems to have been probably the history of the changes during and since Cretaceous times in the centre of the continent is:—

1. The existence of a great marine area in which the Upper Cretaceous rocks forming the Rolling Downs system were deposited, and by which the central district now forming the Higher Steppes and including the McDonnell, James and other ranges in the Centre was isolated from both the botanical Autochthonian region on the west and the comparatively narrow coastal strip on the east and south-east.

2. After the elevation and partial denudation of the Rolling Downs system another submergence occurred, when the same region was occupied partly by a marine but mainly and especially in the central-southern area by a great Lacustrine area, and at this time the Desert Sandstone (Supra-Cretaceous) was deposited.

3. The Lacustrine area gradually diminished, but pluvial conditions or at all events a greater rainfall than the present one continued into Pliocene times.

4. During the latter periods the Coastal Range, then much higher than now, formed a barrier between the large, internal, well-watered area and the narrow coastal strip.

5. In Post-Pliocene times desiccation ensued.

So far as the flora is concerned the original division of the continent into a western and an eastern half, the former containing the Autochthonian constituent, is generally admitted. At the present time, the former, which, as is generally agreed upon, was isolated from the eastern half during Cretaceous times, contains the typical Australian series of genera and is, in this respect, to be strongly contrasted with the Euronotian or eastern flora which was, as Professor Tate says, "superimposed by the Oriental and Andean incursions." To these may be added the same author's conclusion that the Eremian flora was developed in Central Australia in Pliocene times "from Autochthonian and Euronotian elements and largely modified by Oriental immigrants."

In connection with this it may be noted that there is considerable difference of opinion with regard to the existence of a cosmopolitan flora in the sense in which the term is used by Professor Tate when speaking of "that primitive flora which marks the close of the Cretaceous and the early stages of the Tertiary period, as has been made known chiefly by the researches of Baron von Ettingshausen." If such a flora did exist then it is somewhat difficult to understand the relationships of the flora of the Autochthonian region.

The date of the prevalence of this supposed cosmopolitan flora is given by Professor Tate in his general conclusions referring to the Larapintine flora (Botanical Report, p. 135), as Supra-Cretaceous and continuing into Paleocene times. That is, it originated subsequently to the time at which the Cretaceous sea, in which the great Rolling Downs formation was deposited, separated the western island off from the eastern coastal area, and during which time the Autochthonian flora which subsequently spread eastwards was being developed. This Autochthonian flora, which on this supposition antedated the cosmopolitan flora, already contained the now more typical series of Australian forms, the Euronotian having been more modified by Oriental and Andean immigration. If the present typical Australian flora is to be regarded as derived from the Autochthonian, then it is somewhat difficult to see the exact rôle played by a cosmopolitan flora which appeared on the scene *after* the development of the present typical Australian flora.

If it be, on the other hand, suggested that this Autochthonian itself is to be regarded as a part of the cosmopolitan flora,* then it is a somewhat curious fact that in the present western flora, which has been to a very large extent (in the restricted area to which Professor Tate has applied the name of Autochthonian region) shut off by barriers from an immigration of Oriental and Andean types, we only find, and abundantly so, representatives of typical Australian genera and not a trace of such doubtful forms as Quercus, Betula, Salix, etc., upon the presence of which in fossil remains the theory of the cosmopolitan flora in Australia really rests. If the Autochthonian was directly derived from the cosmopolitan flora, then we might surely have expected to find some relics of such genera, and the entire absence of them and the presence amongst endemic genera of only the typical Australian flora of the present day seems to be, so far as it goes, strong evidence against the existence of Baron von Ettingshausen's cosmopolitan flora.

* In this case of course the date of the cosmopolitan flora must be assigned to an earlier period than Supra-Cretaceous and Paleocene or even late Cretaceous as the Autochthonian flora, as Professor Tate says, was "dismembered in Cretaceous times," in fact, during Upper Cretaceous times it was isolated by the sea in which the Rolling Downs formation was deposited.

Whilst this matter is one upon which two contrary opinions are held, there can be no doubt but that Professor Tate's botanical regions, especially taking into account the existence of two subsidiary divisions in his larger Euronotian region, indicate a most important addition to our knowledge of the general features and relationships of the Australian flora.

As Professor Tate stated in his address "On the Influence of Physiographic Changes in the Distribution of Life in Australia," his work had reference mainly to the flora and that in the case of the fauna it yet remained to zoologists "to fuse the species into geographic groups."

Inasmuch as our present knowledge of the Central Fauna is now considerably more complete than it was before the Horn Expedition and that in the case of certain large and important groups such as the Mammalia and Lacertilia we are now better able to judge of the relationship of the fauna of various parts of the continent, it may be worth while both to indicate the general relationships of the fauna of the central area and to attempt to outline certain general faunal regions into which probably the continent may be divided.

In certain respects the fauna stands in strong contrast to the flora. We find no great Autochthonian region occupying the western and south-western part of the continent. There is amongst the higher forms no series of characteristic Australian animals, unless it be to a certain extent amongst the lizards and birds, which can be considered as having been largely represented and developed in this western area during its long period of isolation, in fact amongst mammals it would seem, judging by their present distribution and the almost entire absence of any which may be regarded as at once primitive and peculiar to the west, that the latter did not actually possess any when it first became separated off from the east in the Cretaceous period during the deposition of the Rolling Downs formation.

In Australia we have thus an ancient western flora which contained representatives of the forms upon which the present floral regions are based whilst the same region probably did not contain many representatives of the more highly developed animals upon the present distribution of which faunal regions must be largely based, though at the same time it contained representatives of lower groups which have also to be taken into account, the members of certain higher groups only reaching it at a later period.

Hence it is that the floral and faunal areas of the continent are, in certain important respects, far from being co-incident.

The details with regard to the distribution of the members of the various groups represented in the central fauna have already been given in the summary of zoological work. The general conclusion with regard to each may be stated briefly as follows:—

The Monotremata are represented by one species widely distributed over all the continent except the north-west, the Marsupialia consist of species characteristic of all the interior but not including certain characteristic genera of the north-east and the south-eastern coastal district including Tasmania, the Rodentia are clearly derived from the east, the birds represent in the main a series widely dispersed over the southern half of the continent, the Amphibia are very few in number and are closely allied to eastern species, the lizards represent both ubiquitous, northern and perhaps especially western forms, whilst the Mollusca are on the whole western forms with a slight admixture of eastern and north-eastern but with none of the characteristic forms which have travelled from Tasmania northwards along the east coast, while in the case of Microphyura, amongst the Mollusca and Acanthodrilus, amongst the earthworms we have rare examples of forms which have evidently travelled in from the north-east by way of an ancient land connection, stretching southwards to the east of the present continent—a connection which gave to New Zealand a certain admixture of such Australian types of plants as travelling from the west had reached this portion of the eastern coast.† It may at the same time be taken for granted that there were then no marsupials present in the west or centre or assuredly the path which could be traversed by a Microphyura or Acanthodrilus could also be traversed by a mammal as it was, in all likelihood, by the struthious birds. At this time, which probably coincided with the upheaval of the Rolling Downs formation above the level of the Cretaceous sea, there must have been a means of communication across from the north-east to the centre and away to the west, which is a point of considerable importance in regard to the early distribution of certain now distinctive Australian types.

Speaking generally, there is no evidence pointing to the fact that in the case of the most important groups of Australian animals—the Monotremes and the Marsupials—the old western part of the continent has any claim to the title Autochthonian. If this were so, then we might expect to find, at all events in the well-watered south-western portion, the lower group—the Monotremata—well represented, whereas the Platypus does not extend to West Australia and the Echidna is as widely, in fact more widely, distributed over the eastern portion.

---

† Wallace, "Island Life," 1st Edit., p. 168.

Nor again in the case of the Marsupials do we find any distinctive forms, such as we might have expected to meet with, amongst the polyprotodonts, with the single exception of Myrmecobius, which however extends right across from the inland borders of Queensland and New South Wales to West Australia, and may just as reasonably be regarded as having wandered across from the east to the west as *vice versa*.

It has apparently been sometimes taken for granted that the West Australian fauna contains, as contrasted with the rest of the continent, ancient and primitive forms, but this conclusion is not, at all events so far as the higher forms of life are concerned, borne out by the facts. Amongst the marsupials, for example, the only genera confined to it are to be found amongst the diprotodonts and not, as might have been expected, amongst the polyprotodonts.

What constituted the fauna of the large western area during the time when the Cretaceous sea separated it off from the east we have little means of ascertaining. Amongst the Mollusca it may have been the early ancestors of the Xanthomelon group, a few examples of which passed across the centre to the north-east; amongst the Amphibia the peculiarly Australian genera are eastern forms only comparatively poorly represented in the western fauna and cannot be regarded as having been developed in the west; amongst the Lacertilia perhaps representatives of the family Pygopodidae and of other forms such as Amphibolurus may have existed, but it is difficult to believe that either Monotremes or Marsupials can have been present.

It is quite true that the proportion of Polyprotodont species present in the west as compared with Diprotodont is greater than in the case of Victoria, New South Wales and Queensland, but this is simply due to the fact that these were developed on the eastern side of the continent and thence spread west and south.*

This absence (if the fact be established) from the western area in times preceding the Upper Cretaceous period of the ancestors of the Monotreme and Marsupial fauna is of importance in connection with the probable way in which the latter entered Australia, for there has been no direct land connection between the north-west and Asia since that period. If they were not in the western area when it was dismembered then we are reduced to their reaching Australia by one of two routes (1) *via* an uplifted Torres Straits, and (2) *via* a south-eastern connection with Antarctic lands, and so across to South America. The former route is practically negatived by the feeble development of the polyprotodont fauna in

---

* Detailed evidence in regard to this is given in the author's Presidential Address to the Biology Section, Aust. Ass. Adv. Sci., Hobart, 1890, pp. 118-120, on "The Fauna and Zoological Relationships of Tasmania."

north-eastern Australia, for it has evidently spread northwards rather than southwards along the east coast, and it may also be added that the absence of Platypus in the north-east is evidence against this route of migration, so that we are, in reality, brought to the conclusion that the primitive marsupial, and possibly the primitive Monotreme fauna also, entered Australia from the south.

The discoveries of recent years with regard to the extinct marsupial fauna of South America together with the alliance between Australia and the latter continent as shown by such form as Cystignathous frogs, certain birds and amongst fishes by the Cyclostomata and Galaxias, etc., and Gundlachia amongst Molluscs, point to a former land connection across Antarctic regions.

Apart from the question of an ancient connection of Australia with Asia there must have been two other connections existing :

(1) The first of these was, according to Mr. Wallace, with North-East Australia itself and a land stretching southwards to the east of the continent and now represented by various land-remnants—New Zealand, New Caledonia, Lord Howe and Norfolk Islands—and accounting both for the presence of certain Australian types of plants in the New Zealand flora and also as previously referred to for the presence of Microphyura amongst Molluscs and Acanthodrilus amongst earthworms which are not found in the south-eastern parts of the continent, and probably also for the distribution of struthious birds. Mr. Hedley, on the other hand, is of opinion* that the element in the Australian fauna indicating affinity with New Zealand is to be sought for in the connection of a similar land area with an older Papuan land which was again united to the north-east of Australia.

(2) The second connection was, according to the theory herein advocated, between the south-eastern part of Australia, stretching across what is now Tasmania, and allowed of the introduction of the early mammalian fauna by way of a land connection with South America.

At this time what is now Bass Straits was dry land, allowing of communication with the south-eastern part of the continent, whence animals could spread northwards along the east coast and westwards into the central and southern parts of the continent. The first of these connections probably took place after the elevation of the Rolling Downs (Upper Cretaceous) series, and the second at a somewhat later period and at a time when what is now New Zealand had lost the connection with the southern antarctic lands, by way of which it probably

* Nat. Science, 1893, p. 187.

received such portions of its fauna as Acanthodrilus, while the connection between New Zealand and the north-east of Australia (or the Papuan land) must have disappeared before the marsupial fauna had reached so far north on the continent. This second connection must however have taken place before Pliocene times, as then Australia had a well developed marsupial fauna, and may perhaps have taken place just at the close of the Cretaceous period and before the deposition of the Eocene beds which exist, as at Table Cape, along the northern shore of Tasmania.* Judging by the absence in the latter of certain typical Diprotodonts, as well as of the Dingo, there has been, at any rate, no land connection between Tasmania and the continent during or since the Pliocene period.

If this be so, then at the close of the Cretaceous period, whilst the rich Australian flora was located mainly in the western and south-western part of the continent and was gradually extending over to the east, the main portion of the at present typical Australian fauna, at least so far as the Mammalia, Pisces, Amphibia, and perhaps to a lesser extent the Aves and Reptilia are concerned, was located in the south-east and eastern parts of the continent and was gradually spreading north and west.

A slightly later union across the Torres Straits allowed of a passage further north of certain types amongst the marsupials and a passage south into the continent of other forms, such as the true Rana.

The present fauna may therefore be regarded as consisting of some four elements which may be very briefly outlined as follows:—

(1). An older one derived from a land connection with Asia, the constituents of which it is difficult to define and which existed partly in the western and partly in the eastern division when these two were separated. We may perhaps regard as representatives of this original fauna such forms as Xanthomelon amongst the mollusca of the western area, Peripatus amongst the Arthropods and Ceratodus amongst the fish of the eastern side. It is also quite possible that along with the development of the Autochthonian flora were developed in the western area, such characteristic Australian families of birds as the Meliphagidae (honey eaters) and Trichoglossidae (brush-tongued parakeets), and amongst lizards the well marked Pygopodidae and perhaps others such as the members of the genus Amphibolurus and the curious *Moloch horridus*, which at the present day are characteristic features of the western fauna.

* Or there may even have been a double connection, one in late Cretaceous and one in Miocene times.

(2). A series derived from a connection with a land area now lying to the east of the continent (and connected also with the Papuan region) represented by Microphyura and Acanthodrilus amongst lower forms and the struthious birds amongst vertebrata.

(3). A series derived from the Austro-Malayian region and including such forms as the Paradiseidae and Megapodiidae amongst birds, the true Rana amongst amphibia, lizards such as Heteronota, Physignathus, etc., earthworms such as true Perichaeta, etc.

(4). A large and important series derived from the south and indicating a former connection with South America across Antarctic lands during a period not later than the Miocene. These include, amongst mammalia, the ancestors of the marsupialia, amongst amphibia certain cystignathous frogs, and amongst fishes Aphritis, Haplochiton, Galaxias, and the lamprey Geotria.

Whilst there are considerable difficulties to be met—principally in the way of explaining why certain forms are not present in Australia—if this connection with South America be granted yet it must be allowed that with an increase in our knowledge of the past and present distribution of various forms the evidence in favour of such a connection, as advocated by such writers as Forbes, Beddard and Hedley on various grounds, has steadily increased, and, at the present time, it is difficult to account for the distribution of the marsupials and other forms mentioned, in any other way.

I have endeavoured above to show that the evidence is against the existence of primitive marsupial types in the old western area of the continent when it was separated from the eastern part, while the diminution of polyprotodonts as we pass north along the eastern side is strong evidence against their having entered Australia across the Torres Straits. There has been, further, no direct connection with the Asian continent since the east and west parts of Australia became united in late Cretaceous times, and we are therefore reduced to the supposition that they reached Australia by way of America.

The development and distribution of the existing fauna within the Australian continent has been largely influenced by (1) the condition of the interior, and (2) the existence of a high range running parallel to the south-east and eastern coast lines and separating off a narrow but fertile and well-watered coastal strip of land from a larger internal area, which since Pliocene times has been gradually becoming more and more dry, with the result that a climatic barrier has replaced an earlier one formed by the Cretaceous sea.

If we go back to the close of the Cretaceous period we find that after the elevation of the bed of the old Upper Cretaceous sea, the west and east were probably united and the conditions of climate were such that animal and plant life could spread across.

Following upon this was a period during which a barrier existed in the form partly of a marine, but most largely probably in the form of a great lacustrine area; with a diminution of this area the central part of the continent probably presented a land surface, watered by large rivers, which were fed by an abundant supply of water partly from the ranges in the centre, partly from those fringing the east and south-east coast. The latter were doubtless higher than at present and during the continuance of these pluvial conditions may even have been capped with snow.* At all events they served as a barrier separating a coastal fauna from an internal fauna. At the present time the barrier is a climatic one, dependent upon the difference of rainfall, but at that time as there was an abundant rainfall and consequently an abundant supply of vegetation on both sides of the coastal ranges, the barrier must have been of a different nature from that which exists at the present day and is probably to be found in the then greater height of the ranges.

If now we suppose the marsupial fauna to have entered Australia from the south across what is now Tasmania, there was a period during which the incoming fauna could travel along two routes, one leading up the eastern coast, the other westwards towards what is now South Australia, across the lower country where the present Dividing range sinks away at its western end.

This primitive marsupial fauna consisted of the representatives of polyprotodont forms, which gradually spread over the continent in all directions.

On the eastern side of the continent over what is now the dry interior of New South Wales, South Australia, and Queensland spread a vast tract of country then covered with an abundant vegetation and from which there was in Northern New South Wales and Southern Queensland, to a certain extent, a passage to the eastern coastal district where the ranges were more irregular and less marked than in the south-east.

It is possible that even the early ancestors of the Diprotodonts reached Australia from South America but as yet the evidence in favour of this is very scanty.

---

* In his "Geographical History of Mammals," Mr. Lydekker states that Victoria possesses "a mountain range whose summits are perpetually clothed with snow." This is rather misleading, for though it is just possible that snow may remain through the length of the summer in small patches in very sheltered spots on Mount Kosciusko there is no such thing as a perpetually snow-capped mountain, much less range, in Australia.

However, whether the earliest forms of true Diprotodonts were developed within or outside of the limits of Australia we are probably safe in concluding that the characteristic Diprotodonts of the region were developed in the great, fertile eastern area of the interior and along the coast, and spread thence over the continent, and northwards into the Papuan region.*

The larger forms now extinct, such as species of Diprotodon, Nototherium, Phascolonus, Macropus, Protemnodon, etc., reached their greatest development in Pliocene times and were characteristic of the eastern interior, spreading southwards round the western end of the Dividing Range into Victoria. They do not seem to have reached the eastern coastal district.

In Post-Pliocene times, with the increasing desiccation of the whole central area they became extinct, though this extinction cannot be attributed wholly to the drying up of the land, because in certain parts, such as Western Victoria, to which they reached, the state of desiccation did not supervene; but at the same time it may perhaps be justly argued that the desiccation of the vast area of the interior was the largest factor in their extinction.

Another fact may be noted with regard to this extinct fauna, and that is that at the time of its development communication with Tasmania had apparently been shut off, at any rate no representative of this fauna reached the island, nor did the Dingo, which appeared on the mainland prior to the final disappearance of the large Diprotodontia, as its remains have been described by Sir F. McCoy as occurring in Pliocene deposits at Colac in company with those of Diprotodon.

Side by side probably with the specialisation of the Diprotodonts the less modified Polyprotodonts were likewise giving rise to the existing types, but amongst them no such relatively gigantic forms were developed as amongst the former.

Perameles, Dasyurus, Phascologale and Sminthopsis spread widely over the whole of the continent from Tasmania in the south to New Guinea in the north, and from the Indian coast on the west to the Pacific coast on the east. Certain forms, however, such as Myrmecobius which may perhaps represent a primitive and little modified form, Peragale, Choeropus, Antechinomys, Dasyuroides and the anomalous Notoryctes either not spreading beyond or being gradually confined to the drying up interior of the continent, whilst on the other hand Thylacinus and Sarcophilus must be regarded as south-eastern forms, the immediate ancestors of

---

* This origin of the true Diprotodontia on the eastern side of Australia was suggested in the author's paper on "The Fauna and Zoological Relationships of Tasmania." Aust. Ass. Adv. Sci., vol. iv., Hobart, 1892.

which may perhaps have entered somewhat later than the original polyprotodont fauna. Their range on the continent was restricted as compared with that of other polyprotodonts and this, despite the fact that they were, judged by their strength and ferocity, quite as well able to maintain a footing as their close allies the species of Dasyurus. Possibly they owe their extinction on the mainland to competition with the Dingo with which, being to a large extent arboreal as well as terrestrial, the Dasyurus did not enter into such close competition.

The existing Diprotodonts we may divide into four groups :—

(1). A widespread and presumably early developed series comprising representatives of the genus Macropus, Bettongia, Potorous, Dromicia,* Trichosurus and Pseudochirus which we may regard as having been developed in the great central area, and as having spread thence in all directions.

(2). A series of forms also developed in the central area and confined to this and the south-western parts of the continent. This includes representatives of the genera Petrogale, Onychogale, Lagorchestes, Caloprymnus, Lagostrophus and Tarsipes, some of which are now widely distributed over the area while others are confined to one or more portions.

(3). A series which may be regarded as having been developed in the sub-tropical and tropical portions of the north and especially the north-east, and comprising the genera Dorcopsis, Dendrolagus, Hypsiprymnodon, Distæchurus, Phalanger, Æpyprymnus and probably Petaurus.†

(4). A series which may be regarded as having been developed in the south-eastern district including what is now the coastal parts of Southern New South Wales, Victoria, and also Tasmania, though the latter was separated off from the mainland before the full development of these forms. This series comprises the genera Acrobates, Gymnobelideus, Petauroides, Phascolarctos and Phascolomys.

---

* It may be noted with regard to this genus that whilst it is an old form it is not restricted in its distribution to New Guinea, Western Australia and Tasmania, but certainly occurs on the mainland. Mr. Thomas, in the Brit. Mus. Cat., 1888, p. 140, states in a footnote that he thinks it likely that the specimens of *D. unicolor* (= *D. nana*) described by Krefft as from the neighbourhood of Sydney, had escaped from captivity. During the last few years Mr. Dudley le Souef has captured *D. nana* at Gembrook in Gippsland, and a specimen of the same species has been secured by myself on the Black Spur Range in Victoria and by Mr. A. Purdie at Sale, in Gippsland. There is no reason to think that these specimens have escaped from captivity.

† This genus ought perhaps to be included in the next series, but its distribution over the northern parts of the continent and in New Guinea, would seem to ally it rather with the north-eastern than with the south-eastern series. It is most strongly developed at the present day, in the coastal districts of New South Wales and Victoria, but is not present in Tasmania (except as an introduced form). It was evidently one of the later developed arboreal diproto-donts, as is shown by its absence from Tasmania.

In his recently published work on the "Geographical History of Mammals,"* Mr. Lydekker adopts the more generally accepted theory that the primitive marsupial fauna entered Australia by way of south-east Asia; though he grants the importance of the discovery of dasyuroid marsupials in the Tertiary rocks of Patagonia as pointing towards the existence, at some period, of a direct communication between the south of America and Australia, and points out the importance of the determination by Messrs. David and Smeeth of the nature of the rocks brought back in the recent cruise of the "Antarctic," by Mr. Borchgrevink as indicating a continental area. In fact Mr. Lydekker makes the following important statement:—"It may be observed that it appears impossible to adequately explain the presence of a Notogeic element in the fauna of Neogaea without the aid of some form of southern land connection; although there is not sufficient evidence to show in what latitude such connection (or connections) existed.†

Mr. Lydekker is of opinion that the original immigration of early polyprotodont forms took place across what is now New Guinea and so into north-east Australia, and there "where they have since been isolated from any serious competition with the higher mammals, they flourished and developed to a degree which they could not possibly have attained to in any other part of the world under existing conditions."‡

At the same time Mr. Lydekker grants that the evolution of the Diprotodonts took place within the limits of the Australian continent and that therefore the Cuscuses—the most typical Papuan marsupial—are to be regarded as immigrants.

Now it is the Papuan region firstly and the north-east portion of Australia secondly, which are remarkably poor in Polyprotodonts; such as the Papuan region possesses are confined to three genera, *Dasyurus* (one species), *Perameles* (six species), *Phascologale* (five species), which, it may be remarked, are the most widely distributed of all the Australian Polyprotodonts and the most capable of adapting themselves to the arid climate of the interior, or the more genial coastal climate from cool Tasmania in the south to tropical New Guinea in the north. On the supposition that New Guinea lay in the line of migration of the primitive marsupials it is an inexplicable fact that here where they can and do live in small numbers, free from competition with higher forms, we have still so little trace of Polyprotodonts and not a single form which is not widely dispersed over the continent. The

---
* Cambridge Geographical Series, 1896
† *Loc. cit.*, p. 127.
‡ *Loc. cit.*, p. 61.

poverty of Polyprotodont life can scarcely be attributed to their having been driven south for which there is no reason whatever or to competition (certainly not with Diprotodonts, as in Australia the two groups exist in large numbers side by side), while it can be most naturally explained by the fact that the Papuan was the last and not the first land of the Australian region to be reached by the marsupial fauna.

Mr. Lydekker regards the occurrence of Australian types of Rats in the Phillipines as of "the utmost importance in respect to Australia having received its mammalian fauna from south-eastern Asia,"* though at the same time he states that they must be regarded as comparatively recent immigrants and "are of comparatively small size, so that it is possible that their ancestors may have been introduced without a direct land connection with any other part of the world." There does not appear to be of necessity any connection at all between the line of rodent and that of marsupial immigration ; the fact that the rodents have come down from the north does not appear to prove that the marsupials did, any more than it proves that the fish Galaxias is an immigrant from Asia.

Again Mr. Lydekker says,† after referring to the alliance between Dasyuridæ and the Didelphyidæ, "This being so, it is a fairly safe assumption that both families are descended from a single common ancestral stock which, apart from any question of a connection between Australia and South America, can hardly have originated anywhere than in the northern hemisphere, seeing that the Didelphyidæ are totally unknown in Notogæa," and then he makes the suggestion that the Dasyuridæ and Didelphyidæ were both differentiated in south-eastern Asia, whence " Representatives of the former family soon afterwards found their way into Australia and New Guinea, while the opossums would appear to have dispersed in one direction into Europe and in the other into North America, eventually making their way from the latter country at a late epoch in the Tertiary period into South America."

In respect to this it may be pointed out that there is as yet, as Mr. Lydekker himself says, no evidence of fossil Tertiary marsupials in Asia, and further, that even if the Dasyuridæ and Didelphyidæ are supposed to have developed in that region, the difficulty of accounting for the non-appearance of the latter in the Australian region is still at least as great as, if indeed not greater than, on the supposition that there was a connection between South America and Australia.

The question with regard to the external relations of the present fauna of the Australian region so far as its affinity with that of South America and Australia is concerned may perhaps be briefly summed up somewhat as follows. The principal elements in the fauna, the distribution of which has to be accounted for, can be divided into two groups—(1) a smaller one, which is common to Polynesia, Australia and South America; (2) a larger one, common only to South America and Australia. The former includes forms such as Acanthodrilus, Microphyura and Galaxias. The latter includes forms such as Gundlachia, Aphritis, Haplochiton, Geotria, Cystignathous frogs, and certain closely-allied South American and Australian Marsupials.

It has been suggested that a land connection via Polynesia, between South Australia and Patagonia, would suffice to explain the distribution; but even if we suppose* that "the Polynesian mammals (if they existed) were drowned out by submergence," in which case one might ask what would happen to the other elements of the fauna such as fresh-water fish, land mollusca, earthworms and struthious birds, such a single connection will not suffice to explain matters.

Any such connection via Polynesia was either with the very north east of Australia or with the latter via a Papuan land; but if we take into account the distribution in Australia of the two groups of animals concerned we find that the second and more important group is essentially, except in the case of Cystignathous frogs, a group of south-eastern and Tasmanian forms, whilst neither Acanthodrilus nor Microphyura occur in this part, but are on the contrary essentially north and north-eastern forms.

No single land connection such as the one suggested will serve to account for the facts of distribution and certainly not one via Polynesia.

Assuming, as we are practically now obliged to do, some southern form of land connection between Australia and South America, the history of this may have been somewhat as follows.

Perhaps in late Cretaceous times both what is now Patagonia on the one hand and a southern extension of Australia across Tasmania on the other hand were in connection with the land mass to which Mr. Forbes has given the name of Antarctica. At an early period and certainly before any mammalian life had reached this land a southern extension of the New Zealand land area was, for a short time only, in connection with the same and so gained the elements in its

* cf., Lydekker, "Geographical History of Mammalia," p. 127.

fauna of southern origin. New Zealand was to the north connected directly or indirectly with North east Australia and so passed on to these Acanthodrilus and Microphyura, gaining in return certain animals and plants from the Australian region. It must for example have acquired its Peripatus in some such way. By this Antarctic land after New Zealand had lost its connection with it a primitive polyprotodont fauna passed across to Australia, entering the latter in the south east and thence spreading gradually over the continent and giving rise therein to the various existing types, whilst in America the same primitive group gave rise to the Didelphyidæ.* Across this land there also spread the Cystignathous frogs, fishes such as Galaxias, Aphritis and Haplochiton and amongst molluses, Gundlachia.

The only way in which it seems possible to account at once for the presence of forms such as Prothylacinus in the Patagonian Tertiary beds and the absence of any of the Didelphyidæ in Australia is to suppose that on the South American side the connection between the Antarctic land and what is now Patagonia was lost at a time comparatively soon after the early polyprotodonts had passed across and during which the Didelphyidæ were being developed perhaps in the more northern part of South America. The Antarctic land must however have been in connection with the Australian continent, perhaps it was in the former itself that such forms as Thylacinus were developed ; at all events the distribution of this in Australia points to its having entered by the south-east at a comparatively late period. It was confined in its distribution to the eastern and especially the south-east and did not spread up to the far north or across to the west.

If, subsequently, the Antarctic land became reunited with South America, then this would account for the finding in the Patagonian beds of such forms as Prothylacinus† allied to Australian marsupials, though they apparently became extinct and did not spread far northwards. Before any of the typical American types could pass across the Antarctic land and reach Australia the latter had lost its connection with the former and thus there would exist the Dasyuridæ in Australia and the Didelphyidæ in South America, and in the latter also certain forms closely allied to Australian types of Marsupials.

Probably the Patagonian Cænolestes and its extinct allies stand in regard to their dentition, in much the same relationship to the Polyprotodonts as Mr.

---

* On this supposition the Didelphid type may be regarded as originating in America and spreading thence across to Europe.

† It is perhaps worthy of note that the relatively restricted distribution of Thylacinus and its extinction on the mainland and preservation only in Tasmania is curiously paralleled by the extinction of its ally in Patagonia, and its restriction also to the south of the continent.

Oldfield Thomas has suggested that the Peramelidæ do in respect of their Syndactylous feet and are to be regarded as a group confined to America and not genetically allied to the Australian Diprotodonts.

The remaining groups may be dealt with briefly as follows.

Rodentia.— The rodents, which are doubtless comparatively late immigrants which entered by way of the north, are represented by six genera, of which Hydromys with two species, Mus with twenty-seven species are widely scattered; Xeromys with one species is confined to the north-east, Conilurus with thirteen species is characteristic of the interior region, scarcely being represented in the east and south-east coast; Uromys with two species is confined to the north-east and east coast, and Mastacomys is represented by one species living in Tasmania and fossil in the interior of New South Wales and also by an undetermined species in Central Australia. Its distribution points to its being an early introduced form which has largely become extinct.

Chelonia.—This group is represented by two genera, Emydura (Chelymys) and Chelodina. The former contains three species, of which one, *E. macquariæ*, is limited to the rivers of the interior and does not occur in the east or south-east coastal rivers, while two others, *E. krefftii* and *E. latisternum*, occur in the north east only. The second genus is also represented by three species, of which *Chelodina longicollis* occurs on both sides of the Dividing Range, while *C. expansa* is a Queensland and *C. oblonga* a western and northern form.

Crocodilia.—This group is represented by two species, *Crocodilus porosus* and *C. johnstonii*. The former is a widely distributed species, the latter is confined to Australia and both are characteristic of the north and north-east, to which they are now confined, whilst the extinct form *Palimnarchus pollens* is found fossil in the Lake Eyre district.

Aves.—The most important points in the distribution of birds in Australia so far as the subdivision of the continent into faunal areas is concerned are the following: The Megapodiidæ or mound birds are represented by three genera, Talegalla, Megapodius and Leipoa. Of these the first two are distinctly north-eastern forms, not spreading into the interior or to the south, whilst the single species of the third (*L. ocellata*) is as distinctively a central form spreading right across from the internal parts of New South Wales, Victoria and South Australia to West

Australia, but not passing across the Dividing Range into the south-east or across to Tasmania. The family has evidently come down from the north-east and one species has been modified in accordance with the dry, arid nature of the interior, the others being characteristic of the dense, fertile scrubs of the north-east.

The family Paradiseidæ is one, the members of which evidently wandered in from their home in the Papuan land and are now especially characteristic of the north and north-east. One species of Bower bird (*Chlamydodera maculata*) has passed across to the west and another (*C. guttata*) has accustomed itself to the dry interior.

Just as with the Megapodiidæ and Paradiseidæ the north-east is the central home of the existing struthious birds, though fossil remains in the internal area show that the same period which was characterised by the development of the large Diprotodonts was also the age of large struthious birds now extinct. At the present day the north-east has both Dromaius and Casuarius, while Dromaius is spread over the east of the continent with one species, *D. irroratus*, characteristic of the western and perhaps north-western side.

The interior and west is mainly distinguished by the absence of many genera confined to the eastern and south-eastern coastal districts, amongst which one of the most prominent is the lyre bird (Menura) which does not extend far north into Queensland, and with its three species may be regarded as belonging especially to the south-eastern fauna. As absentees from the large western area may also be noted genera such *Dacelo*, which is in the main a northern form, out of the five species only one coming as far south as Victoria, *Alcyone, Cisticola, Pœphila, Geocichla, Philemon, Ptilinopus*, etc.; other genera such as *Podargus* which is in the main an eastern and northern one, *Graucalus, Collyriocincla, Gerygone, Sericornis, Zosterops*, being but poorly represented.

The genus Amaurodryas is confined to the south-eastern corner of the continent, including Tasmania, while other genera such as Amytis and Acanthiza range widely over the interior, east and south but do not extend into the north-east.

Lacertilia.—In regard to Lacertilia it is possible that some of the Australian genera were modified during the early time when the west and east were separated. At the present day there are twenty-two genera endemic in Australia, and thirteen exotic to Australia. Amongst the latter *Gymnodactylus* is widely distributed over the world, *Phyllodactylus* occurs in Tropical America, Africa and the Mediterranean islands, *Varanus* occurs in the Moluccas, Celebes, New Guinea, Polynesia, Africa and South Asia, *Lygosoma* (= *Lygosoma* + *Hinulia* + *Emoa* + *Siaphos* + *Rhodona* of various authors) occurs in Australia, East Indies, China, Tropical and South Africa, *Ablepharus* in South and West Asia, South-east Europe, Tropical and South Africa.

A second series occur less widely distributed, but in such parts that they may be regarded as having entered the continent from the north. These are — *Ceramodactylus* found in Arabia and Persia ; *Thecadactylus* in the East Indies, *Gehyra* in the islands of the Indian and Pacific Oceans and on the west coast of Mexico, *Lepidodactylus* in the East Indies and Polynesia, *Gonyocephalus* in the East Indies etc., Nicobar, and Andaman Islands, *Physignathus* in Polynesia, Siam and Cochin China, *Tiliqua* in Australasia and the Indo-Malayan Islands. Lastly, of exotic genera there remains *Ebenavia*, previously known only from Madagascar and closely allied to Phyllodactylus, from which it differs in the absence of claws on the digits. At the present day Phyllodactylus, as a modification of which Ebenavia may be regarded, occurs in Western Australia, and it is at any rate possible that the Central Australian Ebenavia and the Madagascar forms are independent modifications of a widely-spread type.

All the species of the genera are endemic in the Australian regions except *Ablepharus boutonii*, which is "irregularly distributed over the hotter parts of both hemispheres."—(Brit. Mus. Cat., Vol. IV., p. 346).

The endemic genera may be divided into three main groups :—

1. A northern and north eastern group (or if not confined to this area most largely represented therein) represented by *Rhynchedura*, *Heteronota*, *Œdura*, *Chlamydosaurus*, *Hemispheriodon*.

2. A western, central and southern group, represented by *Diplodactylus*, *Nephrurus*, *Amphibolurus*, *Moloch*, *Tympanocryptis*, *Ophidiocephalus*, *Chelosauria*.

3. A series of widely-spread genera represented by *Pygopus*, *Delma*, *Lialis*, *Egernia*, *Trachysaurus*.

Amongst the first of the groups it is noticeable that there has been a certain amount of passage from the north across the north west and so to the west and *vice versa*, whilst certain genera, such as *Amphibolurus*, are in the main characteristic of one region, though they have certain species, such as in that genus *A. barbatus*, widely distributed over the continent.

The lizards of the south-east part of the continent do not form in any way a distinctive group and are represented partly by forms widely distributed over the whole continent, partly by eastern coastal species, partly by western and southern, but with the exception of Physignathus, they do not contain a representative of the distinctly northern group.

Taken altogether Australia does not show any marked affinity in the matter of lizards with any other region, its exotic genera having probably entered by the north-east and there is little or no affinity with South America. Speaking of the distribution of lizards Wallace says,* " on the whole the distribution of the Lacertilia shows a remarkable amount of specialization in each of the great tropical regions, whence we may infer that Southern Asia, Tropical Africa, Australia and South America each obtained their original stock of this order at very remote periods, and that there has since been very little intercommunication between them." The absence of any marked affinity between Australia and other regions in the Lacertilian fauna stands in marked contrast to that of other orders, but in connection with this it may be pointed out that the climate of any Antarctic land connection, though temperate enough to suit mammals, may not have been favourable to the migration of such heat-loving creatures as lizards. It may indeed be said that we find an alliance existing between Australia and South America amongst the groups in the case of which we might have expected to do so if that alliance be due to a connection across a moderately cool Antarctic land.

Amphibia.—In the Amphibia one family, the Cystignathidæ, is common to Australia and South America. Out of fifteen genera no fewer than twelve are endemic. Rana is represented by one species in the Cape York peninsula, and this genus together with Hyla may be regarded as an immigrant from the north. The stronghold of the endemic genera is undoubtedly the eastern and south eastern coastal district and though some of the more widely dispersed species may perhaps represent forms once more widely distributed but now separated by the gradual desiccation of the interior, others doubtlessly owe their wide dispersal to the remarkable power which they possess of accommodating themselves by burrowing and storing up water to districts which they can only traverse during irregularly recurring rainy seasons.

Pisces.—Amongst fresh-water fishes the more important forms are :—

(1). Ceratodus, the remnant of a more widely dispersed form, now confined to the Mary and Burnet Rivers in Queensland, though fossil remains found in the

* Distribution of Animals. Vol. II., p. 401.

Lake Eyre district show that it was formerly more widely scattered over the continent.

(2). Osteoglossum, a genus represented in tropical Australia and South America.

(3). A series showing a strong affinity with South America and confined in Australia to the south-eastern part of the continent and Tasmania (not passing or scarcely at all north of the Dividing Range in Victoria) and including Haplochiton, Aphritis and the species of Galaxias, and amongst Cyclostomata, Geotria.

So far as the distribution of forms in the Australian region is concerned the most important point is the clear demarcation of a south-eastern series from an interior series* which is especially characteristic of the Murray River system. The former includes Lates Microperca, Aphritis, Haplochiton, Prototroctes, Galaxias, Agonostoma, Gadopsis, and amongst Cyclostomata Geotria and Mordacia. The latter includes Oligorus, Therapon, Murrayia, Ctenolates, Chatoessus, Copidoglanis.

Vermes.—(*a*) *Oligochaeta.*† Amongst Earthworms three families are represented in Australia, viz., Perichaetidae, Cryptodrilidae and Acanthodrilidae. At present our knowledge with regard to the Earthworm fauna is limited almost entirely to the eastern side of the continent, though we may feel sure that it will be found to be much more abundant here than in the drier west.

Certain forms of Lumbricidae are very common (Allolobophora and Allurus), but as their range is restricted to the neighbourhood of settlement, in which they have almost completely ousted the indigenous fauna, they may safely be regarded as introduced forms.

On the eastern side the following are the more important points in regard to distribution :—

Acanthodrilus is entirely confined to the north, not reaching further south than Queensland and having one species, doubtless derived from the north-east, in the Centre. True Perichaetes are also confined to the north.

Diporochaeta is represented by two species in North Queensland, but is otherwise restricted to the south-east (six species). Fletcherodrilus has one species in Queensland and one in New South Wales.

---

* Details with regard to this are given in a paper by Mr. A. H. S. Lucas "On the Vertebrate Fauna of Victoria," read before the Royal Society of Victoria, in July, 1896, and now in course of publication in Proc. R.S. Vict., vol. ix.; and also in the Presidential Address of the author to the Biology Section of the Australasian Association, Hobart, 1892, "On the Fauna and Zoological Relationships of Tasmania."

† The classification followed is that given by Beddard in his Monograph of the order Oligochaeta, 1895.

The more southern portion of the coastal district is characterised by the presence of Cryptodrilus, Digaster, Megascolides (the stronghold of which is Victoria), Trinephrus, whilst Megascolex and Diporochaeta are most largely developed here, though to a certain extent they spread northwards.

Victoria together with Tasmania, coastal New South Wales and Queensland, may be regarded at present as possessing three groups of Earthworms, the Queensland and northern fauna being marked off from the other two by the presence of Acanthodrilus and true Perichaeta and the entire absence of Megascolides and to a very marked degree of Megascolex, Diporochaeta and Cryptodrilus

(*b*) *Turbellaria*.—At the present time, though a considerable number of land planarians have been described, it is scarcely possible to say anything very definite with regard to their distribution in lack of any knowledge of their occurrence over the whole of the west and north-west and to a very large extent the north east of the continent. Geoplana is widely distributed over Tasmania, New South Wales, and Queensland and extends across to New Zealand. In regard to the species present in the latter Dr. Dendy,* to whom we are indebted for the greater part of our information relative to the Australasian forms other than those of New South Wales, which have been described by Messrs. Fletcher and Hamilton, says : " I find that few of these (twenty species) can with safety be absolutely identified with Australian species, yet in several cases the differences are extremely slight and not such as would, in my opinion, justify a specific distinction if the varieties were found together." The genus Rhynchodemus so far as at present known is present in Australia in Victoria, increases in number in New South Wales and is recorded also from Queensland, but only doubtfully from New Zealand, while Cotyloplana is only known from Lord Howe Island.

On the continent the genus Geoplana is the dominant one, but in curious contrast to the close similarity of certain species of New Zealand and Victoria those of the latter and New South Wales seem, with the exception of a very small proportion, to be distinct.

(*c*) *Nemertinea*.—So far as at present known the only Australian species are *Geonemertes australiensis* and *G. novæ-hollandiæ*. On the continent the former is as yet known only from the eastern and south eastern part, including New South Wales, Victoria and Tasmania.

In the three latter groups, viz., Oligochaeta, Turbellaria and Nemertinea, the fauna of the eastern coastal district stands as might have been expected in strong contrast to that of the extensive but dry internal area of the continent.

* Aust. Ass. Adv. Sci., Brisbane, vol. vi 1895, p 118.

Crustacea.—Amongst land and fresh-water forms the more important are the following :—Lepidurus is found in the eastern coastal district from New South Wales through Victoria to Tasmania and reaches westwards into the coastal parts of South Australia. The internal area of the continent, from the inland parts of Queensland, New South Wales and Victoria on the east side right across to West Australia, is characterised by the absence of Lepidurus and the presence of Apus.

Amongst Brachyura the land crab *Telphusa transversa* is a north-eastern form which has spread across to the Centre. Amongst the Macrura one form (*Astacopsis bicarinatus*) extends over practically the whole of the continent, occurring right in the Centre and all along the eastern and south-eastern coast from Queensland to Victoria. Tasmania is distinguished by the presence of distinct species (*A. franklinii*), but the most important fact is the presence of a genus of burrowing land crayfish, *Engæus*, confined to and characteristic of the south-east and Tasmania, and not even extending so far north along the east coast as New South Wales.

Mollusca.—In the land and freshwater molluscs we can distinguish (1) a north eastern or Queensland group. With regard to this Mr. Cooke* says, "The strip of coast-line from Cape York to the Clarence River stands apart from the rest of Australia, and is closely connected with New Guinea. There can be little doubt that is has been colonised from the latter country, since an elevation of even ten fathoms would create a wide bridge between the two. Many of the genera are quite strange to the rest of Australia." In this area the more important genera are *Hadra*, which here reaches its maxium, *Rhytida*, *Chloritis*, *Planispira*, *Panda*, *Thersites*, *Stenogyra*, amongst slugs *Janella* and amongst fresh-water forms *Isidora*.

It is from this region that the species of *Hadra*, *Chloritis*, *Planispira*, *Thersites*, *Stenogyra* and the ancestors of *Isidorella* must have passed across to the centre where they have since been isolated. This would appear to show that the Rhytididæ, which are now established in the north and of which, as Mr. Hedley says, "originating in Antarctica," one group "established itself in Tasmania and marched in force to Cape York and even crossed to Mount Owen Stanley in New Guinea," had not reached far enough north in time to allow them to pass across to the centre.

(2). A restricted West Australian group represented by species of *Liparus*, *Pupa*, *Succinea*, and the group *Rhagada* amongst Helices.

---

* "Molluscs and Brachipods." Cambridge Nat. Hist., p. 322. The information with regard to the distribution of the Mollusca referred to, except those of Central Australia, is taken from this work.

(3). A series belonging to eastern and south-eastern Australia and Tasmania. Our knowledge of Victorian land mollusca is at present unfortunately very imperfect, but as in other groups a fuller knowledge will doubtless reveal a more or less close alliance between those of southern Victoria and Tasmania.

In this area the species of *Hadra* diminish from the north to the south, none being known in Tasmania; *Cystopelta*, *Carvedes*, and *Helicarion* are common to the mainland and Tasmania, while the slug *Aneitea graeffei* is common to New South Wales and Queensland. The last operculate, a Helicina, is found in the north of New South Wales none being present in Tasmania, so that the true south eastern part of the continent is devoid of these.

FAUNAL DIVISIONS OF THE AUSTRALIAN REGION.

In his "Geographical History of Mammals," Mr. Lydekker has adopted the name Notogæic Realm to include the Australian, Polynesian, Hawaiian and Austro-Malayan Regions. The Australian Region includes Australia, Tasmania, New Guinea, and the adjacent Papuan Islands.

So far as the distribution of animals and plants is concerned, in regard to the Australian region, we have to deal with a series of events which may be briefly summed up as follows:—

(1). A division in late Cretaceous times of the land area into an eastern and a western portion.

(2). A union of these two divisions and the final formation, at all events in the southern-central part of the continent, of a great lacustrine area accompanied by more or less pluvial conditions, and resulting during Tertiary times in the existence of a vast internal area, of which Lake Eyre may be regarded as the centre, suitable for the development of animal life.

(3). During this period the eastern and south-eastern coastal range, then probably of much greater height than at present, formed a barrier between (*a*) the eastern coastal lands, and (*b*) those of the interior and west.

(4). A land connection (*a*) across Torres Straits with a Papuan area and either directly or indirectly with the Polynesian region, and (*b*) one across Tasmania stretching southwards to an Antarctic land and so allowing of communication with the Næogæic Realm (South America).

(5). The obliteration of these two land connections and the final isolation of the Australian continent.

(6). A gradual drying up of the interior, the physical barrier of the coastal ranges being partly replaced, partly intensified, by a climatic barrier dependent upon the dryness of the interior and the humidity of the coastal region.

The result of these series of events was the division of the continent into two main areas:—

(1). A northern, eastern and south-eastern coastal land coinciding in this part with the present rainfall limit of 25-50 inches per annum.

(2). A large central, western and southern area comprising the rest of the continent.

Owing, in the first instance, to the northern connection with Papua (and also Polynesia) and to its southern connection with an Antarctic land and, to a lesser extent, to differences in temperature, the first of these areas contains two well-marked faunas.

(a) A north and north-eastern.
(b) A south-eastern and south.

The north-eastern area may be regarded as closely united with Papua and we can thus at the present time divide the Australian Region into three sub-regions which may be distinguished as follows:—

(1). The **Torresian** sub-region. This includes Papua and north and north-eastern Australia as far south as the Clarence River. On its north-western side it merges as might be expected to a certain extent into the western area. It is characterised by such forms as Proechidna, Dorcopsis, Dendrolagus, Hypsiprymnodon, Phalanger and Distoechurus; Xeromys amongst Rodents; Casuarius, Megapodius, Talegalla, and the Paradiseidæ amongst birds; Rhyncœdura, Oedura and Physignathus amongst lizards; Crocodilus amongst Reptilia; Rana amongst Amphibia; Ceratodus and Osteoglossum amongst fishes; Acanthodrilus and true Perichætes amongst earthworms; Microphyura, Hadra, Chloritis, Janella, Isidora, etc., amongst land and fresh-water mollusca.

The name *Papuan* has already been suggested by Mr. Hedley for this sub-region but the name *Torresian* is here suggested both as being less liable to lead to confusion and as suggestive of the position of the old land connection which gave rise to the faunal affinity of its now separated northern and southern parts.

(2). The **Bassian** sub-region. This includes the eastern and south-eastern coastal strip, lying between the coast line and the Dividing Range south of the

Clarence River, and also Tasmania. On the mainland it naturally merges to a certain extent, where the Dividing Ranges falls away at its western end, with the fauna of the interior but in the main it is strikingly dissimilar to this.

It is characterised by such forms as Acrobates, Gymnobelideus, Petauroides, Phascolarctos, Phascolomys, Thylacinus and Sarcophilus (the two latter now confined to Tasmania) amongst mammals; Amaurodryas amongst birds; Myxophyes, Philocryphus, Phanerotis and the strong development of Lymnodynastes amongst Amphibia; Lates, Microperca, Girella, Aphritis, Agonostoma, Gadopsis, Prototroctes, Galaxias, Mordacia, Geotria amongst fishes; Gundlachia, Cystopelta, Helicarion amongst Mollusca; Diporochaeta, Cryptodrilus, Digaster, Megascolex and Megascolides amongst Earthworms. The most important feature of the fauna is the South American affinity.

The name is adopted from that of Bass Strait, across which, when uplifted, the South American continent must have passed.

(3). The **Eyrean** sub-region. This includes the whole of the interior, southern and western part of the continent, the coastal ranges on the east and south-east separating it from the Torresian sub-region in the north-east and the Bassian region in the south-east.

It is characterised by such forms as Myrmecobius, Notoryctes, Dasyuroides, Antechinomys, Petrogale, Onychogale, Lagorchestes, Caloprymnus, Lagostrophus, Tarsipes amongst Marsupialia; Conilurus (Hapalotis) amongst Rodents; Megaderma amongst Bats; Leipoa amongst Birds; Diplodactylus, Nephrurus, Amphibolurus (most largely), Moloch, Tympanocryptis, Ophidiocephalus, Chelosauria amongst Lizards; Emydura amongst Chelonians; Myobatrachus amongst Amphibia; Oligorus, Ctenolates, Murrayia, Copidoglanis, Plotosus amongst Fishes; Apus, Eulimnadia and the most distinctive types of Estheria and Limnadopsis amongst Crustacea; Liparus and the Rhagada group of Helices amongst Molluscs.

The name *Eyrean* is suggested for this region in consequence of the fact that Lake Eyre is to be regarded as the centre of the great internal Lacustrine region, which was closely associated with the former development of the now extinct series of gigantic Diprotodont and Struthious forms, which during Pliocene times formed perhaps the most distinctive fauna of the continent, whilst it was probably in the area centering around Lake Eyre and during the period of its gradual desiccation that the present fauna of the interior of the continent was developed.

These three faunal sub-regions are indicated on the accompanying map.

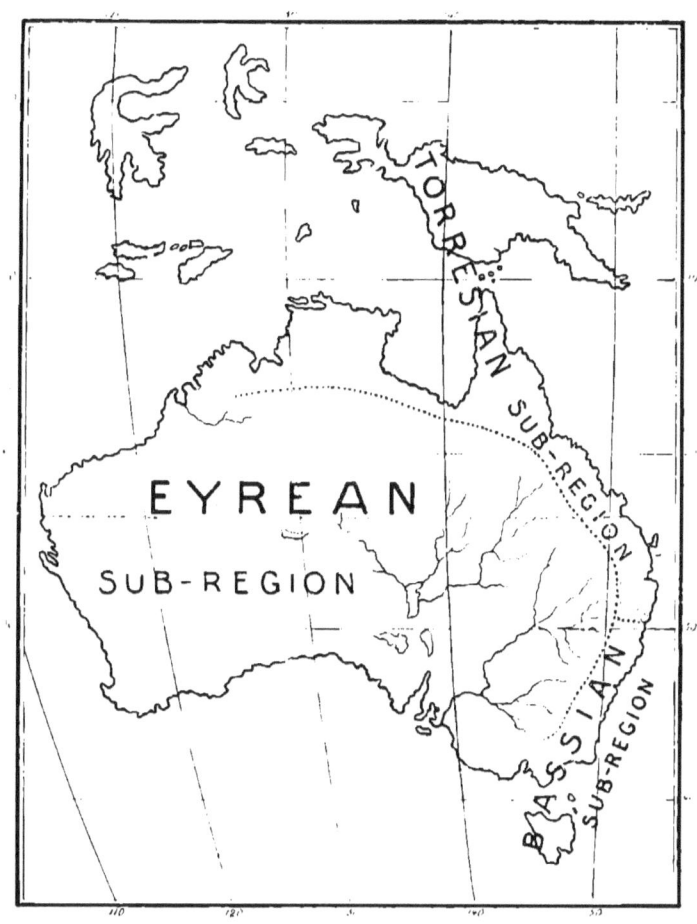

FAUNAL SUB-REGIONS OF THE AUSTRALIAN REGION.

SUPPLEMENT

TO THE

ZOOLOGICAL REPORT.

TABLE OF CONTENTS.

HYMENOPTERA. By W. F. Kirby, F.L.S., F.E.S., Assistant in the
  Zoological Department, British Museum (Natural History)   203
ADDITIONS TO THE FAUNA   210
  (*a*) MARSUPIALIA.
  (*b*) LACERTILIA.

HYMENOPTERA.

By W. F. KIRBY, F.L.S., F.E.S., Assistant in the Zoological Department, British Museum (Natural History).

The collection submitted to me consisted of a number of specimens preserved in spirits and contained in small phials, a large proportion of the specimens being *Formicidæ* (ants). It should, however, be pointed out that, although spirit is a convenient method for collecting specimens, it is undesirable to employ it for any insects except hard-shelled beetles or bugs; for the exposed wings of insects are very liable to get torn in it, and the hair of bees, etc., gets matted together and spoiled.

With the exception of a few specimens, which were too much damaged for identification, a full list of the species obtained is given below. Altogether twenty-eight species of *Hymenoptera aculeata* are here enumerated, of which six appear to be new to science.

HYMENOPTERA ACULEATA.

FORMICIDÆ.

FORMICINÆ.

1. Camponotus schencki.

Camponotus schencki, Mayr, Verh. Zool. Bot. Ges. Wien, XII., p. 674 (1862).

Paisley Bluff (one specimen).

2. Camponotus impavidus.

Camponotus impavidus, Forel., Ann. Soc. Ent. Belg. XXXVIII., p. 155 (1893).

McDonnell Range, under stones on hill-side (several specimens).

3. Camponotus arcuatus.

Camponotus arcuatus, Mayr, Journ. Mus. Godeffroy, IV. (Heft 12), p. 8 (1876).

Hugh Creek, McDonnell Range, July 11, 1894.

Two specimens, apparently belonging to this rare species.

4. Camponotus reticulatus, sp.n.

Length.—Large worker, 9 mm.; small worker, 6 mm.

Large worker.—Dark pitchy-brown, inclining to rufotestaceous; mandibles ferruginous, antennæ, tarsi, and under surface of legs reddish; abdomen with a white stripe on each side intersecting the white incisions; mandibles about twice as long as broad at the base, gradually curved, pointed at the tip, and armed with six large teeth, in addition to the long terminal tooth; clypeus carinated, about as broad as long, the sides subrotund, the upper and lower extremities concave. Outer antennal ridges slightly waved, but diverging above, and neither these nor the central one attain the summit of the vertex. Head very convex behind, thorax sloping, gradually narrowed behind; thorax and abdomen sparsely clothed with thick, raised hairs. Antennæ and legs clothed with short hair; legs moderately long and slender, with a very strong, pale, terminal spine on the tibia. Petiole large, conical, sloping slightly forwards.

The small workers are nearly black, with the scape of the antennæ and the tarsi rufotestaceous, and the incisions of the abdomen pale.

Paisley Bluff, burrow-nest under stones, many specimens; also Palm Creek and Finke Gorge.

I cannot make this conspicuous species agree with any of the specimens or descriptions before me, though it somewhat resembles *C. testaceipes*, Smith. It is possibly a honey ant, but the carinated clypeus is alone sufficient to separate it from *C. inflatus*, Lubbock.

5. Camponotus novæ-hollandiæ.

Camponotus novæ-hollandiæ, Mayr, Verh. Zool. Bot. Ges. Wien, XX., p. 939 (1871).

McDonnell Range; Palm Creek; Paisley Bluff.

Forms burrows under stones on hill-sides; sometimes found solitary.

Many specimens, a variable species; some of the small workers are wholly pale yellow; the large workers have black heads, and their abdomen is reddish-brown with pale incisions, and the under-surface pale.

6. Camponotus denticulatus, sp.n.

Worker.—Length, 9 mm. Black, the antennæ, mandibles, and adjacent part of the face, as well as the thorax and legs, more or less ferruginous; head, body and legs with oblique, short, white bristles; abdomen with a fine, silky pubescence in addition. Head long, mandibles with six rather oblique teeth, the first small, and of nearly equal size; the last apicals are much larger. Clypeus only slightly carinated; antennal ridges strongly marked; thorax gradually sloping, somewhat narrower behind; petiole rounded above, legs long.

McDonnell Range. Ant from mound-nest with a slit opening at the top. Burrows underground. Several specimens. A considerably smaller specimen from Paisley Bluff may also belong to this species.

Appears to be related to *C. novæ-hollandiæ*, Mayr.

7. Camponotus horni, sp.n.

Worker.—Length, 9 mm. Rufous, with a slight purplish suffusion, legs and petiole purplish above, abdomen with purple and coppery reflections, tarsi rufous. Head smooth, rounded, short; clypeus short, not carinated; antennal ridges not strongly marked, but with a third between them. Scape of antennæ with short, raised bristles. The hairs on the head and body short and erect; those on the legs oblique. Mesothorax much depressed; prothorax and metathorax much rounded above.

Females.—Length 13 lines. Black, shining, with short white bristles, antennæ rufous, legs entirely testaceous, the tibia and tarsi a little darker than the coxæ and femora. Wings smoky hyaline, the fore wings with the crossing narrower, united for a short space at their point of junction.

Palm Creek.

Burrow-nest under stones. Several specimens. The peculiar structure of this species will probably ultimately necessitate its removal to another genus; but the rufous body and purple abdomen will render it easily recognisable.

8. Hoplomyrmus micans.

Polyrhachis micans, Mayr., Journ. Mus., Godeffroy, IV. (Heft 12), p. 21 (1876).

Storm Creek (four specimens).

As the name *Polyrhachis* is preoccupied, I prefer to use *Hoplomyrmus*, Gerst., for this genus.

9. Hypoclinea flavipes, sp.n.

Worker.—Length, 2 mm. Black, very closely and finely punctured, the large metanotum ending in an open crescent, with moderately long, diverging horns; the central ridge also ends in a projection; scale very long, rounded, and flattened; tarsi and more or less of the mouth-parts and antennæ yellow and testaceous.

Ants from Porcupine Grass (*Triodia pungens*) Tempe Downs.

A very small but well-marked species, apparently resembling the much larger *H. scabrida*, Roger, in colour.

PONERINÆ.

10. Bothroponera denticulata, sp.n.

Worker.—Length, 12 mm. Black, pubescent, the mandibles, the extreme tip of the antennæ, the under-surface of the legs, and the incisions of the abdomen more or less rufous. Head and thorax very closely and irregularly rugose and granulated, the pronotum and petiole showing a tendency towards longitudinal striation, head with two strong ridges between the antennæ, and the occiput somewhat concave. Antennæ pubescent, rather stout. Mandibles broad, strongly punctured, sub-triangular, with nine teeth, the second, fourth and sixth smaller than the others; the three last broad, and successively increasing in length; the seventh and eighth with a short notch on the inside at the base. On the other side there are only eight teeth, and the small notches are not visible. The face is set with long yellowish bristles, and the mandibles are also bordered with smaller bristles, of which there is a row above the teeth, which sometimes renders their examination difficult. Tibia with several terminal spines, the last serrated, tarsi set with numerous short spines, as well as hairs. Petiole longitudinally ridged, the ridges terminating in strong teeth behind; the middle one is slenderer and rather longer than the others, of which there are about four on each side.

Blood Creek, several specimens. Allied to the Indian *B. rufipes*, Jerdon.

11. Myrmecia nigriceps.

Myrmecia nigriceps, Mayr., Verh. Zool. Bot. Ges. Wien, XIV., pp. 725-728 (1862).

Reedy Hole, Bagot Creek, Alice Springs (one specimen from each); Ayers Rock and Illamurta (several specimens from each).

MYRMECINÆ.

12. Pheidole longiceps.

Pheidole longiceps, Mayr., Journ. Mus. Godeffroy, IV. (Heft 12), p. 51 (1876).

Paisley Bluff, in burrow-nest under stones.

MUTILLIDÆ.

13. Mutilla rugicollis.

Mutilla rugicollis, Westw., Arcana Entomologica, II., p. 17, Plate 53, Fig. 5 (1844).

Tempe Downs (one specimen).

THYNNIDÆ.

14. Thynnus ochrocephalus.

Thynnus ochrocephalus, Smith, Trans. Ent. Soc. Lond., 1868, p. 231.

Camp, Illamurta (one specimen). A very fine, and apparently rather scarce species.

15. Thynnus obscurus.

Thynnus obscurus, Klug., Abhand. Akad., Berlin, 1812, p. 22, Fig. 1.

Palm Creek (one specimen).

16. Thynnus carbonarius.

Thynnus (Thynnoides) carbonarius, Smith, Cat. Hym. Ins. B.M., VII., p. 23 n. 51 (1859).

One specimen, without locality.

17. Rhagigaster illustris, sp.n.

Male.—Length, 11 mm. Black, legs and apex of the abdomen red, mouth-parts mostly pale yellow; prothorax narrowly bordered with pale yellow before and behind; middle of the scutellum and hind border of the post-scutellum pale yellow, and five pale yellow spots on each side of the black part of the abdomen above, the first linear, the others slightly indented on the outer part of the front edge.

Crown Point (one damaged specimen).

Differs from *R. hæmorrhoidalis*, Guér., in the pale yellow (inclining to ivory white) markings on the abdomen.

SCOLIIDÆ.

18. Scolia læviceps.

Scolia læviceps, Kirby, Trans. Ent. Soc. London, 1889, p. 117.

Adminga Creek (one specimen).

19. Campsomeris radula.

Tiphia radula, Fabr., Syst. Ent., p. 354, n. 5 (1775).

Alice Springs (three specimens); George Gill Range (two specimens).

BEMBICIDÆ.

20. Bembex raptor.

Bembex raptor, Smith, Cat. Hym. Ins. B.M., IV., p. 326, n. 40 (1856).

Crown Point (one specimen).

POMPILIDÆ.

21. Pompilus morio.

Sphex morio, Fabr., Syst. Ent., p. 349, n. 16 (1775).

Storm Creek (one specimen).

22. Pompilus semiluctuosus.

Pompilus semiluctuosus, Smith, Cat. Hym. Ins. B.M., III., p. 166, n. 231 (1855).

Rudall Creek, Gosse Range (one specimen).

23. Agenia fusiformis.

Agenia fusiformis, Sauss, Reise der Novara, Hymenoptera, p. 53 (1867).

Opossum Creek (three specimens); Darwent Creek (one specimen).

SPHEGIDÆ.

24. Sphex canescens.

Sphex canescens, Smith, Cat. Hym. Ins. B.M., IV., p. 246, n. 37 (1856).

Crown Point (one specimen).

25. Sphex luctuosa.

Sphex luctuosa, Smith, Cat. Hym. Ins. B.M., IV., p. 250, n. 47 (1856).

Alice Springs (two specimens); Dalhousie (one specimen).

The *Sphegidæ* in the collection are injured by spirit, rendering their identification somewhat uncertain.

VESPIDÆ.

EUMENINÆ.

26. Eumenes latreillei.

Eumenes Latreillei, Sauss., Mon. Guêpes Sol., p. 51, Plate 10, Fig. 5 (1852).

Alice Springs (many specimens); near Idracowra (one specimen); Illamurta (one specimen).

27. Abispa ephippium.

Vespa ephippium, Fabr., Syst. Ent., p. 362, n. 2 (1775).

Bagot Creek (one specimen). A slight variety, differing from the type in having the thorax almost entirely red, instead of black.

28. Odynerus sanguinolartus.

Odynerus sanguinolartus, Sauss, Mon. Guêpes, Sol., Suppl., p. 224 (1854).

Darwent Creek (one specimen).

ADDITIONS TO THE FAUNA.

The following additional species have been received since the publication of the Zoological Report (Part II.) and are included in the table giving the summary of results (Part I., p. 140):—

MARSUPIALIA.

Dasyuridæ.

1. **Dasyurus geoffroyi**, Gould.

Locality.—Crown Point and Alice Springs.

2. **Perameles obesula**, Shaw

Locality.—Alice Springs.

3. **Perameles eremiana**,* Spencer.

Locality.—Charlotte Waters.

4. **Peragale minor**,* Spencer.

Locality.—Charlotte Waters.

OPHIDIA.

1. **Typhlops polygrammicus**, Schlegel.

Locality.—Alice Springs and Charlotte Waters.

LACERTILIA.

Geckonidæ.

1. **Ceramodactylus damæus**, L. and F.

Locality.—Several specimens have been received from the Centre and one from Mr. Dudley le Souëf, collected by him in North Queensland.

2. **Diplodactylus tesselatus**, Günth.

Locality.—Charlotte Waters.

---

* Described in Proc. R.S. Vict. (New Series), vol. ix.

3. **Diplodactylus byrnei,** L. and F.

Locality.—Charlotte Waters.

4. **Diplodactylus conspicillatus,**[*] L. and F.

Locality.—Charlotte Waters.

5. **Œdura tyroni,** De Vis.

Locality.—Alice Springs.

6. **Œdura marmorata,** Gray.

Locality.—Alice Springs.

Pygopodidæ.

7. **Ophidiocephalus tæniatus,**[*] L. and F.

Locality.—Charlotte Waters.

Scincidæ.

8. **Ablepharus greyi,** Gray.

Locality.—Alice Springs.

9. **Ablepharus elegans,** Gray.

Locality.—Alice Springs.

One specimen, which agrees with the type except that there are eighteen longitudinal rows of scales, six supraciliaries and all the dorsal scales have a central black dot.

[*] Described in Proc. R.S. Vict. (New Series), vol. ix.

ALPHABETICAL INDEX TO PART I. NARRATIVE, SUMMARY, ETC.

Abispa ephippium, 209.
Ablepharus greyi, 117, 211, *burtoni*, 117, *ruficaudatus*, red tail of, 26, *elegans*, 211.
Aborigines, see Natives.
Acacia aneura, 13, 122, 126, 161, *cyperophylla*, 13, 16, 161, *homalophylla*, 13, 23, 161, *farnesiana*, 15, *ulicina*, 28, *salicina*, 44, 46, 93, variations in form of, 121, *dictyophleba*, 99.
Acanthodrilus, 180, 187, 193, *eremius*, 63.
Acanthophis antarctica, deaf adder, 42.
Actinoceras tatei, 166.
Adaptation of plants to climate, 15.
Adminga Creek, 22.
Ægialitis nigrifrons, 16.
Æstivation of Frogs, Snails, etc., 21.
Agenia fusiformis, 208.
Alberga River, 15.
Alexandra, Princess, Parakeet, 100, habits and sporadic appearance of, 101.
Alice Springs, 130, et seqq.
Alpita, tails of *Peragale lagotis*, 92.
Amadeus basin, 74.
Amadeus Lake, 74, crossing and camp at, 83, 95.
Amaurodryas, 190.
Amera, or spear-thrower, 38.
Amphibia, 18, remarks on colouration of, 26, rate of growth of, 19, of Central Australia, 149, distribution in Australia, 192.
Amphibolurus, 180.
Amphibolurus pictus, colouration of in relation to protection, 25, *Amphibolurus reticulatus*, 109, 47.
Amphibolurus maculatus, difference in colouration in male and female, 25.
Amphibolurus barbatus, 28, 41.
Ancylus australicus, 97.
Ancitea, 196.
Angasella argicerens, 109.
Antarctica, 187.
Antarctic land, connection between Australia and America, 180, 188.

Antechinomys, habits of, 76, scarcity of, 121.
Antiarra, a ceremonial stone of the Natives, 58.
Ants, 24, Porcupine grass ants, habits of, 69-71, ant nests, 126.
Ant lions, crater and tracks of, 29.
Aphritis, 181, 193.
Apus, 18, 20, 154, 195.
Araneidæ, 157.
Artesian bores, 30, derivation of water supply in Central area, 167.
Arundo phragmites, 74.
Arunta tribe, 36, 39.
Asaphus spp., 171.
Aspidites melanocephalus, 109.
Astacopsis bicarinatus, distribution and habits of, 60, 125.
Atriplex rhagodioides, mealy secretion on leaf of, 13.
Australia, connection of with South America, 179.
Autochthonian region, Botanical, 159, 172, not coterminous with a corresponding zoological region, 176, 177.
Aves, 146, 189, distribution of Central Australian, 146.
Ayers Rock, 85-90.

Bagot Creek, increase of water at, during dry season, 96.
Barriers to migration in the Australian Region, 174, 181, 182.
Bassia, 14.
Bassian sub-region, characteristics of, 197.
Bembex raptor, 208.
Bettongia lesueuri, 28, 84.
Birds, Chestnut-eared finches, 13, dotterels, 16, Black cockatoo, 31, 128, Mysterious appearance and disappearance of, 18, rock pigeons, 99, *Xerophila nigricincta*, 120, superb warblers, 120, cat bird, 120, wedge-tailed eagles, 123, mo-pokes, 124.

Bathinia australis, tenacity of life of, 16, 65.
Blackburn, Rev. T., on the Coleoptera of Central Australia, 156.
Blacks, see Natives.
Blood drawing, 37, 58.
Bony Bream, 54.
Boomerangs, 39.
Bore, at Oodnadatta, Succession of strata passed through, 166.
Botanical sub-divisions of the Australian continent, 172.
Botany, Summary of, 159.
Bothroponera denticulata, 206.
Brachysema, 66.
Brehm, on seasons of Steppe lands, 9.
Brinkley Bluff, 124, 128.
Bull-roarer, 35.
Burrowing Frogs, two kinds of, 43.
Burt Plains, 103, 128.
Butterflies, 41.

Cænolestes, 188.
Callabonna, Lake, 55, 57.
Callitris verrucosa, 57.
Calyptorhynchus stellulatus, 31.
Camel buggy, 5.
Camels, habits of, 3, 4, feeding on thorny plants, 15.
Camel, riding, 7.
Camponotus inflatus, 88, *cowlei*, 88, *midas*, 88, *denticulatus*, 126, 203, *arcuatus*, 203, *horni*, 204, *impavidus*, 203, *novæ-hollandiæ*, 203, *reticulatus*, 203, *schenki*, 203.
Campsomeris radula, 208.
Capparis spinosa, 119, *mitchelli*, 119.
Carabidæ, prevalence of certain species, 125.
Carmichael Crag, 75.
Caryodes, 196.
Case-moths, caterpillars of, and remarks on various kinds of, 44.
Cassias, 13, growth of, 46, flowering, 117, 119, 124, 126.
Castle Hill, 49.
Casuarina Decaisneana, 15, 49.
Casuarius, 190.
Caterpillars, social, 44.
Central Eremian Region, 161.

Ceramodactylus damæus, 148, 210.
Ceratodus, 192.
Ceremonies, sacred, of the natives, 36.
Cernuatia, 118.
Chambers Pillar, 48, myth relating to origin of, 50.
Chandler Range, 57.
Charlotte Waters, 23.
Chatoëssus, 193, *horni*, 54, 65, 105, 115.
Cheiroleptes platycephalus, 18, 21, 26.
Chelifer, 119.
Chelodina, 189.
Chelonia, Australian, 189.
Chelymys, see Emydura.
Chenopodium, 17.
Chignons, worn by men, 92.
Chlamydodera, 190.
Chloritis, 195, *squamulosa*, 116.
Chæropus castanotis, habits of, 109.
Churiña, sacred sticks and stones, 35.
Class divisions of Natives, 36.
Clay-pans, description of, 17, Fauna of 18-22.
Claytonia, 15, 38.
Climate, of Steppes, 9, adaptation of plants to, 15, as influencing animal life, 66.
Coccidæ and ants, 70.
Coglin Creek, 23.
Coleoptera, 117, 156.
Colouration, remarks on with regard to protection, 25, main conclusions with regard to, 27, Alexandra Parrakeet, 100.
Competition amongst plants, 14.
Conglomerate, Post-Ordovician, 166.
Conilurus, 189, *pedunculatus*, 130, 144.
Conlin Lagoon, 132.
Copidoglanis, 193.
Conway, Mount, 124.
Corbicula sublævigata, 170.
Corrobborees, general remarks on, 35, preparation for, 37, at Tempe Downs, 72.
Cosmopolitan flora, Mr. Deane's remarks on, 160, discussion of, 175.
Crabs, freshwater, 16, 21.
Crayfish, 60.
Cretaceous period, Successive changes in Centre of Continent during the, 174.
Crinum flaccidum, 14, 18.

Crocodilia, 189.
Crown Point, 33.
Crustacea, of clay-pans, preference for muddy water, 20, Periodicity in occurrence of and dominance of certain forms, 133, 136, Distribution and affinities of Central Australian forms, 154-156, distribution of, in Australia, 195.
Cryptobius mastersi, 117.
Cryptodrilus, 193.
Ctenolates, 193.
Cunningham Gap, 33.
Curculios, resemblance to bark, 26, predominance of amongst beetles, 125.
Cycads, 77, 113.
Cystignathidæ, 192.
Cystopelta, 196.

Dalhousie, 16.
Danais petilia, 41.
Daniel, Mount, strata forming, 167.
Darling lily, 11.
Dasyuroid marsupials in Patagonian Tertiaries, 185.
Dasyurus geoffroyi, 210.
Davenport Creek, 102.
Deane, H., on a Cosmopolitan flora, 160.
Decoration of person for corrobboree, 35, 37.
Desert country, 9, 78.
Desert gum tree, 81.
Desert Oak, switch-like structure of leaf-stalks, 15, 49.
Desert Sandstones, 16, 167, fossils of, 168.
Development, necessity of rapid, amongst animals, 143.
Didelphyidæ, 186, absence of in Australia, 188.
Didymosurus gleichenioides, 168.
Digaster, 194.
Dingos, in Kamaran's Well, 82.
Diplodactylus byrnei, 211, *tesselatus*, 210, *conspicillatus*, 211.
Diporochæta, 193.
Diprotodon, 55, extinction of, 183.
Diprotodontia, development in Eastern Australia, 183, division of existing ones into four groups, 184.

Distribution, sporadic, of plants, 99, 113, 160, of molluscs, 113, 116.
Dromaius, 190.
Dromicia, distribution of and occurrence in Victoria, 184.
Dryness of the country, 22, 66.
Duboisia Hopwoodi, the Pituri plant, 81.

Earthworms, 63, 109.
Ebenavia horni, 118.
Echidna, 146.
Egernia whitii, variation in colouration of, 26.
Eleotris larapintæ, 105, 115.
Ellery Creek, 121.
Emu, tracked by natives, 76.
Emydura (Chelymys), 189.
Encephalartos Macdonnelli, 77, 114.
Endodonta planorbulina, 116.
Engæus, 195.
Eremian Region, 159, 172.
Eremian Flora, constituent elements of, 161.
Eremophilas, 13, 46, 126.
Estheria dictyon, 116, *packardi*, persistence of, 20, 133.
Estherias, 20.
Ettingshausen, Baron von, suggested cosmopolitan flora, 160, 175.
Eucalyptus rostrata, 13, 33, 58, *oleosa*, 121, *gamophylla*, 59, *microtheca*, 13, 33, 44, 161, *eudesmoides*, 81, *terminalis*, 125.
Euonmus latreilli, 209.
Euronotian region, 159, 172.
Extinct marsupials, 55.
Eyre, Lake, in dry and wet season, 12.
Eyrean sub-region, characteristics of, 198.

Fat tails of Marsupials, 130.
Fauna of Australia, elements constituting the, 180, of Central Australia, 110 *et seqq.*, 177, permanent and fluctuating, 141, affinity with that of South America, 179.
Faunal Divisions of Australian Region, 196.
Faunal Regions of Australia, as suggested by Mr. Hedley, 172.
Ferns, at Reedy Creek. 74.

Ficus orbicularis, 119.
Ficus platypoda, 57, 119.
Finke River, 32, named by Stuart, 32, drainage area of, 32, Gorge, 108.
Fish, 50, 52, 54, 66, 67, 69, 105, 115, 150, 193.
Fletcherodrilus, 193.
Flies, a pest, 24.
Floods, 34, 50.
Flora, constituent elements of Central, 161.
Formicidae, 203.
Frogs, of clay-pans, 18, rapid development of, 19, burrowing and water-holding, 21, at Ayers Rock, 89.
Furina ramsayi, 149.

Galaxias, 181, 187, 193.
Gap, Emily, Temple bar, Simpson, etc., 132.
Gastrolobium, a poison plant, 93, 161.
General Conclusions, 171.
Geology, summary of, 162-171.
Geonemertes, 194.
Geoplana, 194.
George Gill Range, 73.
Geotria, 181, 93.
Gibber plains at sunset, 17.
Gibbers, 11, origin of, 12.
Giddea, 23.
Gillen, Mount, 131.
Glen Helen Station, 100.
Glossopteris, 68.
Gobius eremius, 52.
Goodenia horniana, sporadic distribution, of, 99.
Gorges, formation of, 104, 107, 165, in neighbourhood of Alice Springs, 131.
Gosse Range, 99, 123.
Goyder River, Camp at 29.
Grasshoppers, 24.
Grass trees, 98.
Grevillea agrifolia, 119, 125.
Growth of plants, must be rapid if they are to survive, 14.
Gryllotalpa coarctata, 69.
Gum creeks, 74.
Gum trees, red, 33, swamp, 33, infested by social caterpillars, 44, mallee, 16, 59.
Hadra, 195.
Hall and Pritchard, on the age of plant bearing beds in Victoria, 169.

Hapalotis mitchelli, 75.
Haplochiton, 193.
Heavitree Gap, 131.
Hedley, C., on Central Australian Mol lusca, 153, on Faunal Regions of Australia, 172.
Heleioporus pictus, 18, 89.
Helicarion, 196.
Helichrysum, 47, 49.
Helicina, 196.
Hemiptera, 110.
Henbury, 55.
Hermann, Mount, 112.
Hermannsburg, 111.
Heteronota, 181.
Hibbertia glaberrima, 75, 119, 161.
Hirudinea 110.
Higher Steppes, 9, 62 79, 102-136.
Holoprymnus micans, 204.
Honey ants, nest and habit of, 87-89.
Hoplocephalus stirlingi, 119.
Horn Valley, 100, 103, 106, 123.
Horseshoe-bend, 13.
Hornea pulchella, 149.
Hugh River, 123.
Hydromys, 188.
Hydrophilus albipes, water beetle, tenacity of life, 22.
Hyla gilleni, 150.
Hyla aurea, rate of development of eggs of, 19.
Hyla rubella, 19, colouration of, 26, 125.
Hypoctinea flavipes, 69, 158, 206.
Hymenoptera, 158, 203.

Idracowra, 18.
Illamurta, 61.
Illara waterhole, 67.
Ilpilla Creek, 61.
Irri-akurn, 38.
Irula, sacred sticks, 35.
Isidorella newcombi, ground mud oper culum of, 22, 65.
Isopoda, 110.

James Range, 61, 92.
Janella, 195.
Jerboa rats, 75.
Johnston Range, 54.

Kamaran's Well, 82.
Kangaroos, 93.
King Creek, 81.
Krichauff Range, 112.
Kurtitina, a native well, 85, 95.

Lacinularia, 140.
Lacertilia, Central Australian, 146, distribution in Australia, 190, see also Lizards, division of endemic genera into three groups, 191.
Lagorchestes conspicillatus, 109.
Land, possession of by Natives, 10.
Larapinta, native name for Finke, 32.
Larapinta Land, 2.
Larapintine flora, 159.
Larapintine region, 159.
Lates, 193.
Latrodectes scelio, 157.
Leipoa ocellata, 83, 189.
Lepidurus, 151, 195.
Lepidoptera, 156.
Leschenhaultia divaricata, resinous material from root of, 29.
Levi Range, 73, 165.
Lialis burtoni, 117.
Lilla Creek, 12.
Limnadopsis squirei, 133 *tatei*, 133.
Limnodynastes ornatus, a burrowing frog, 12, 117.
Liparus, 195, *spenceri*, 113.
Livistona Mariæ, 113, 164.
Lizards, difference in colour of male and female, general remarks on colouration, 25, varying susceptibility with regard to heat, 28, at Palm Creek, 117, division of Central Australian forms into groups, 146, variations in, 147.
Loamy plains, 12.
Lophophaps leucogaster, 99.
Loranthus, 17.
Lower Steppes, 9, 11-61.
Lowland vegetation, 160.
Lucas, A. H. S., on the vertebrate fauna of Victoria, 193 (footnote).
Lumbricidæ, introduced into Australia, 193.
Luritcha Tribe, 72.

Lydekker, R., on entrance of marsupial fauna into Australia, 185, Rodents, 186, origin of Dasyuridae and Didelphyidae in South-East Asia, 186.
Macropus rufus, distribution of, 93.
Maiden, J. H., on resins, 71.
Macumba River, 15.
Marsilea quadrifolia, 18.
Mammalia of Central Australia, 140, 143.
Mammals, difficulty of obtaining, 109.
Malurus melanotus, 120, *leucopterus*, 120, *lamberti*, 120.
Mammalia, probable absence of, in the west when the latter dismembered from the east, 176, 177.
Marsupials, extinct, 55-57, severely handicapped when in competition with rodents by having to carry young in the pouch, 127, division of Central Australian forms into three groups, 144, absence from west whilst the latter dismembered from the eastern part of the continent, 178, derivation of Australian forms, 185-189, path of distribution of in Australia, 182.
Mastacomys, 144, 189.
McDonnell Ranges, 102-136.
Megaderma gigas, 143.
Megapodiidæ, 181, distribution of in Australia, 189.
Megascolides, 194.
Melania balonnensis, 69, *venustula*. 170, *lutosa*, 170.
Members of the Expedition, 2.
Menura, distribution of, 190.
Mereenie Bluff, 109.
Meridian ants, 129.
Metura elongata, case of caterpillar, 15.
Microperca, 193.
Microphyura hemiclausa, 64, 187.
Missionary Plains, 98.
Mollusca, survival of *Bithinia australis*, 16, 63, remarks on persistence in Central Australia, 65, at Reedy Creek, 75, Palmer River, 97, Finke Gorge, 109, Palm Creek, 113, 115, distribution and affinities of Central Australian forms, 150-154, distribution of in Australia, 195.

Moloch horridus, 41, 180.
Monotremata, 116.
Mosquitos, 24.
Mound birds, 83.
Mound springs, 16, 82.
Mourning, women in, 39.
Mulga, 13, 81, variations in foliage of, 121, 122.
Munyeru, 15, grinding of, 38.
Murray lily, 14.
Murray River system, fish of, 193.
Murrayia, 193.
Mus, 189, gouldi, 75, hermannsburgensis, 120, musculus, 114.
Musical instruments of Natives, 72.
Mussel, fresh-water, 21.
Mutilla rugicollis, 207.
Myrmecobius, distribution of, 178.
Myrmeleon, 28.
Myriapoda, at Palm Creek, 118, 140.
Myrmecia nigriceps, 206.

Naias major, 60, 119.
Nardoo, 18.
Narcotics, 66, 82.
Native cooking, 91.
Natives, camp of, at Crown Point and general remarks with regard to, 34, height of 39. Camp at Henbury, 58, drawings at Reedy Creek, 78, at Ayers Rock, 90, at Mount Olga, 92.
Necklace containing dead man's hair, 59.
Nematocentris tatei, 69, 105, winneckei, 105, 115.
Nemertinea, 194.
Nephila eremiana, 59.
New Guinea, poverty of Polyprotodonts, 185.
New Zealand, relation of to southern land, 187.
Nicotianum suaveolens, 66.
Ninox boobook, 124.
Notoryctes, remarks on, 52-54, 115.

Obsidian bombs, suggested origin of, 170.
Odynerus sanguinolartus, 209.
Œdura marmorata, 211, tryoni, 211.
Old man porcupine, 97.

Olga, Mount, 90.
Oligochæta, 158, distribution of genera in Australia, 193.
Oligorus, 193.
Oodnadatta, 5.
Ophidia, 149.
Ophidiocephalus tæniatus, 147, 211.
Orange, native, 119.
Ordovician strata, 163, area of 164, fossils, 164, constitution of, 164, folding, 165, correlation of, 171.
Orthis dichotomalis, 166.
Orthis leviensis, 171.
Osteoglossum, 193.

Paisley Bluff, 124.
Palœarea wattii, 166.
Palæontological results, Summary of, 170.
Pallimnarchus pollens, 170, 189.
Palm Creek, 113-120, fauna of, 115-119.
Palmer River, 67, 97.
Palorchestes azael, 55.
Panda, 195.
Papuan Region, Polyprodont fauna of, 185, as suggested by Mr. C. Hedley, 173, 197.
Paradiseidæ, 181, 190.
Peragale, species, native names and habits of, 110, distribution of, 145.
Peragale lagotis, 34, 109, minor. 115, 210.
Perameles eremiana, 210, obesula, 210.
Perichæta, 181, 193.
Petermann Pound, 96.
Petermann Creek, 73, 165.
Petrogale lateralis, 77, 126.
Photographing, difficulties of, 80.
Phratries, division of Tribe into, 36.
Phascologale, 84, macdonnellensis, 130, cristicauda, 130.
Pheidole longiceps, 207.
Philonthus subcingulatus, 117.
Phlogius crassipes, 131, stridulating organ of, 135.
Phyllodactylus, 148.
Phyllopoda, 154.
Physignathus, 181.
Physignathus longirostris, habits of, 30.
Pine Point, 99.

Pisces of Central Australia, 150, of Murray River, 150, distribution of in Australia, 192, as showing South American affinity, 193; see also *Fish*.
Pitchis of Natives, 39.
Pituri plant, uses of, 81.
Planarian, water, 74.
Planispira, 195.
Plants not crowded together, 14, relation to animals, and climatic environment, 15, dominance of certain genera, 126.
Plotosus argenteus, 66.
Plumbago zeilanica, 131.
Polyprotodonts, path by which the primitive forms entered Australia, 179, Lydekker's views, 185.
Polynesia, connection of with Australia, 179, 187.
Polytelis, see Spathopterus.
Polyzoa, 21, 110.
Pomatostomus rubeculus, 120.
Pompilius morio, 208, *semiluctuosus*, 208.
Porcupine grass, 59, description and figures of, 81, 97, 100, 126.
Porcupine grass ant, habits of, 69-71.
Portulacca, 15.
Post-Ordovician Conglomerate, 166.
Potamogeton Tepperi, 119.
Pre-Cambrian formation, 102, 162, foliation planes of, 163, evidence as to age of, 163.
Prickly plants, 14.
Property, ideas of amongst the natives, 40.
Protective colouration, remarks on with regard to lizards, etc., 25.
Prothylacinus, 188.
Prototroctes, 193.
Pseudonaja affinis, 75.
Ptilotis leucotis, 63.
Ptilotus, 14, 47.
Pupa, 195, *ficulnea*, 116.
Pygopodidae, 180.
Pyrameis cardui, 11.

Rain season, change in fauna, 112.
Rana, 192.
Rabbit-Bandicoot, 34, 109.
Redbank Creek, 103.
Redbank Gorge, 104.
Red gum, 1, 30, 33.

Red Mulga, 13, deciduous bark of, 16.
Reedy Creek, description of camp on, 71.
Resin, derived from Porcupine grass and used by ants in making nests, 71, use by natives, 39, 71.
Rhagada, 195.
Rhagigaster illustris, 207.
Rhodona bipes, 117.
Rhynchodemus, 191.
Rhytididae, 195.
Rivers, 31.
Rock pigeons, 99.
Rock wallabies, 126.
River gum, 13.
Rodentia, Australian, 189.
Rodents, habits of mice and Jerboa rats, 75, Mus spp., 120, of Central Australia, 111, migrations of, 111.
Rolling Downs formation of Queensland, 166.
Roly Poly, 13.
Rotifers, 21, 110.
Rudall Creek, 99.
Running Waters, 60.

Salsolaceous vegetation, 17.
Salt-bush, 13.
Salsola kali, 13, 17.
Sandhills, 18, 19.
Sarcophilus, 183.
Saxatile Plants, 119.
Saxatile vegetation, 160.
Scelia laeviceps, 208.
Scorpions, 11, 118.
Scrub, description of, 13, 16.
Seasons, of Steppe Lands, 9.
Secular changes in Australian continent, as influencing distribution, 174.
Shield of natives, 39.
Silicification of Upper and Supra-Cretaceous rocks, 169.
Skink lizards, susceptibility to heat, 28.
Sminthopsis psammophilus, 81, *crassicaudata*, 121.
Snakes, 42.
Sonder, Mount, 102.
South America, affinity between fauna of and that of Australia, 187, *et seqq*.
Spathopterus alexandrae, 100, 116.
Spears, 38.

Sphex canescens, 209, *luctuosa*, 209.
Spiders, large orb-web of *Nephile eremiana*, 59, at Palm Creek, 119.
Spinifex grass, 11, 31.
Stenogyra, 195.
Steppes, Australian, 8, 9, division of into Lower and Higher, 9.
Stevenson River, 15.
Stinking acacia, 13, 23.
Stridulating organ in spiders, 135.
Struggle for existence, certain conditions of in Central Australia amongst animals and plants, 112.
Styphelia mitchelli, 106, 161.
Succinea, 195.
Succulent plants, prevalence of in dry region, 15.
Summary of Results, 139.
Supra-Cretaceous formation, 167.
Sutherland, A., experiments on rate of development of eggs of *Hyla aurea* at different temperatures, 19.
Sarsinsonia canescens, sporadic distribution of, 99, 161.
Swamp gums, 13, 33.

Tachys spenceri, 97, 117.
Taeniopygia castanotis, 13.
Talegalla, 189.
Telphusa transversa, 21, 155, 195.
Tempe Downs, 68.
Temperature at nights, 97.
Tertiary formation, 170.
Therapon, 193, *percoides*, 67, 105, 115, *truttaceus*, 105, 115, *fasciatus*, 105.
Thersites, 195, *adcockiana*, 63.
Thorns, no protection against camels, 15, adaptation to climatic environment, 15.
Thylacinus, 183, extinction of on mainland, 188, footnote.
Thynnus carbonarius, 207, *obscurus*, 207, *ochrocephalus*, 207.
Tietken's camping ground at Mount Olga, 91.
Tiliqua occipitalis, killed by heat of sand, 28.

Tobacco plant, 67.
Torresian sub-region, characteristics of, 197.
Totems, traces of in Arunta tribe, 58.
Tracking, of natives, 76.
Tribulus, one of the prickly plants, 11.
Trichoglossidae, 180.
Triodia, 85, *pungens*, 71, 85.
Trinephrus, 194.
Trora, native musical instrument, 73.
Tubaija, 110.
Turbellaria, 194.
Typhlops polygrammicus, 210.

Unio stuarti, 21.
Unterjatta, of a native well, 82.
Upper Cretaceous formation, 166.

Varanus gilleni, colour of, 25.
Variations in structure of body dependent upon seasons as shown by size and number of teats in certain species of Marsupials, 143.
Vermes, 193.
Vermicella annulata, burrowing in sand, 42.

Walker River, 68.
Wallace, on the Australian region, 172.
Water, from root of Mallee gum, etc., 22.
Water-holes, 13, 15, 51, 91, 109.
Water holding frog, 21.
Water-plants, 60, 65, 74, 119.
Waterhouse Range, 123.
Weapons of natives, 38.
Winnall's Ridge, 81.

Xanthomelon, 153, 180.
Xanthorrhoea Thorntoni, 98, 161.
Xeronnys, 189.
Xerophila nigricincta, 120.
Yarrumpa, native name for honey ant, 88.

Zeil, Mount, 100.
Zoology, summary of, 139-158.

www.ingramcontent.com/pod-product-compliance
Lightning Source LLC
Chambersburg PA
CBHW021352230426
43666CB00006B/499